OTHER BOOKS BY H. E. HUNTLEY

Dimensional Analysis, Macdonald & Co. Ltd., London
(Reprinted by Dover Publications, Inc.).

Nuclear Species, Macmillan & Co., Ltd., London.

The Faith of a Physicist, Geoffrey Bles, London.

THE DIVINE PROPORTION

Radiograph of the shell of the chambered nautilus (*Nautilus pompilius*), a beautiful and well-known modern sea shell. The successive chambers of the nautilus are built on a framework of a logarithmic spiral. As the shell grows the size of the chambers increases but their shape remains unaltered (courtesy Kodak Limited, London).

THE DIVINE PROPORTION

A Study in Mathematical Beauty

by H. E. HUNTLEY

DOVER PUBLICATIONS, INC., NEW YORK

Published in Canada by General Publishing Company, Limited,
30 Lesmill Road, Don Mills, Toronto, Ontario.
Published in the United Kingdom by Constable and Company, Ltd.,
10 Orange Street, London WC2.

The Divine Proportion: a Study in Mathematical Beauty is a
new work, first published in 1970 by Dover Publications, Inc.

Standard Book Number: 486–22254-3
Library of Congress Catalog Card Number: 70–93195

Manufactured in the United States of America
Dover Publications, Inc.
180 Varick Street
New York, N.Y. 10014

Preface

Having taught mathematics and physics to students of university standard for thirty years, I find myself still facing the didactic difficulty that confronted me when I first began to teach, *viz.*, how to instill in adolescent minds the best, and perhaps the only enduring motive for reading mathematics, which, in my view, is the aesthetic motive. = of beauty → art

Like the improvement of a bodily skill, the appreciation of beauty in any form can be developed with practice. This involves presenting the learner with objects of beauty for his appraisal; in the present instance with a collection of mathematical specimens of acknowledged aesthetic appeal. Anthologies of poetry or of music are not hard to come by: *Gems from Shakespeare* or *Collected Folk Songs* are familiar titles, so why should not a comparable collection be made for mathematics, which after all is an art form? To the aesthetically minded mathematician much mathematics reads like poetry. Then let the following collection be read as an anthology, which, in the Greek derivation, is a collection of flowers. As far as I know it is the first of its kind.

The topics discussed are simple; they are all more or less directly connected with the "golden section" of the Greeks. The author hopes that the essay will not only make an aesthetic appeal to the reader, but will also stimulate his own creative activities, for the experience of creating something new or of uncovering some hidden beauty is one of the most intense joys that the human mind can experience.

I am glad to express my gratitude to my friend, Charles Clutter-buck of Bath University, for his skilful execution of the drawings and to Mr. Peter Kurz for reading the manuscript and for many helpful suggestions.

Hutton,   H. E. HUNTLEY
Somerset,
U. K.

C O N T E N T S

Beauty is a conspicuous element in the abstract completeness aimed at in the higher mathematics; it is the goal of physics as it seeks to construe the order of the universe; it ought at least to be the inspiration of all study of life. . . . It raises for us the question of the depth and reach of our awareness. This needs the poet's prayer that more of reverence in us dwell.

JOHN OMAN, *The Natural and the Supernatural*

Euclid alone has looked on Beauty bare . . .

EDNA ST. VINCENT MILLAY, *Sonnet*

Introduction

The theme of this book is the aesthetic appreciation of mathematics. Poincaré's remark that " . . . but for harmony beautiful to contemplate, science would not be worth following" is applicable also to his own discipline, mathematics. K. Weierstrass's dictum that "No mathematician can be a complete mathematician unless he is also something of a poet," recalls Poincaré's: "The mathematician does not study pure mathematics because it is useful; he studies it because he delights in it and he delights in it because it is beautiful." Since the purpose of these pages is didactic, it has *intended for instruction* been necessary to consider how the reader's emotion could be stirred as he exercised his intelligence in following a mathematical argument; how a mathematical idea could be made to be both convincing by its logic and moving by its beauty. Three steps seem to be required, which, though they may be necessary, may not in themselves always be sufficient, to kindle the spark of aesthetic feeling into a flame.

First, if we seek to implant in the budding mathematician a feeling for beauty in the topics which he studies, we must confront him with beautiful specimens. No argument would convince a blind man of the beauty of a rainbow; he must see it. Accordingly,

I

in the following pages is presented a short "anthology," a selection of mathematical specimens which experience shows to have aesthetic appeal. The field of choice, however, is so wide that a strict rule of limitation is expedient. It would not be too difficult to range widely and write a sort of mathematical belles-lettres, but the alternative was preferred and it will be found that all the specimens selected are related to a single idea, one that appealed strongly to the aesthetic sensibilities of the ancient Greeks from the days of Pythagoras, *viz.*, the "divine proportion" and the related topic, the Fibonacci series. The basic idea is simple; most of the cognate topics are within the competence of high school pupils. The inference the reader may be expected to make is that, if so restricted a field contains so many gems, the whole realm of mathematics must be rich indeed!

Second, some preliminary education related to the selected specimens is needed. A limited sense of aesthetic appreciation is given; the rest must be acquired. For example, the mathematically uneducated can easily appreciate the dual symmetry of an ellipse; that is given. But the unlimited store of beauty of the conic sections is reserved for the mathematically trained: it is acquired. This indicates that the path to real aesthetic pleasure is through toil, a principle that holds far beyond the realm of mathematics. Spade work is essential: *per ardua ad astra.*

Third, the neophyte must be encouraged to help himself. The Socratic method is best here, and the reason is simple. The appreciation of beauty is scarcely to be distinguished from the activity of creation. "In the moment of appreciation we ... re-enact the creative act, and we ourselves make the discovery again."[1] Self-help is the royal road to intuitional glimpses of truth and to discoveries which, even at secondhand, bring with them the joy of creative activity.

To summarize: to induce aesthetic pleasure, select a suitable object, acquire the relevant education and help yourself.

EXERCISE OF A SKILL

There are of course other sources of pleasure in the pursuit of mathematics than the appreciation of beauty. There is, for instance, the exercise of mental skills. Mathematics is a language, and skill in its use can afford great satisfaction. J. Bronowski writes:

> Mathematics is in the first place a language in which we discuss those parts of the real world which can be described by numbers or by similar relations of order. But with the workaday business of translating the facts into this language there naturally goes, in those who are good at it, a pleasure in the activity itself. They find the language richer than its bare content; what is translated comes to mean less to them than the logic and the style of saying it; and from these overtones grows mathematics as a literature in its own right. Mathematics in this sense is a form of poetry, which has the same relation to the prose of practical mathematics as poetry has to prose in any other language. The element of poetry, the delight in exploring the medium for its own sake, is an essential ingredient in the creative process.[2]

Another secondary source of pleasure in the pursuit of mathematics is the sense of increased power which accompanies it. "Give me a fulcrum," said Archimedes concerning the lever, "and I will move the world!" Students experience great satisfaction when they first learn to use logarithms, or when the remainder theorem enables them to factor an algebraic polynomial.

But although the acquisition of skill in mathematics and the sense of increased mathematical power contribute to the enjoyment of mathematical pursuits, neither can exceed the joy of the creation of beauty, remembering that even appreciation is a re-enactment of creative activity, so that creating new mathematics and reading old mathematics produced by someone else result in very similar types of aesthetic feeling.

A CAREER IN MATHEMATICS

If you, the reader, contemplate a career in pure or applied mathematics, whether in industry or research, or in the teaching profession, you should be warned that although there can be one infallible, enduring reward for you in this pursuit—joy in creative activity—there stand certain discouraging hazards, of which four may be noted briefly:

1. The burden of hard mental concentration is a *sine qua non.* You may find that you have to live with a problem day and night

for weeks, giving all you have of mental resources in order to solve it: no inspiration without perspiration.

2. Your best efforts may be fruitless. Despite extravagant expenditure of time and skill, the result is nil. Disappointment, frustration and near-despair are common experiences of serious mathematicians.

3. You may be lonely. Scarcely anyone will appreciate your work because few will be capable of understanding it.

4. The results you do obtain will always appear to be disproportionately meager in comparison with the effort you expended to produce them: "The mountain laboured and brought forth a mouse."

The one sure path to satisfaction in a mathematics career is to cultivate assiduously the aesthetic appreciation of the discipline. That pleasure will not fade, it will grow with exercise.

TEACHING MATHEMATICS

As a mathematics student and teacher of long standing I may be permitted to say something about two sorts of students, the pedestrians and the high fliers.

First, teachers should be sympathetic with the mathematically ungifted. For humane reasons, such should be dissuaded from any unaccountable ambition they may have to take advanced mathematics. C. G. Jung has some wise advice for us under this heading:

There are however, others who are by no means unamenable to education, who, on the contrary, exhibit special aptitudes, but of a very peculiar and one-sided nature. The most frequent of such peculiarities is the incapacity to understand any form of mathematics that is not expressed in concrete numbers. For this reason higher mathematics ought always to be optional in schools, since the development of the capacity for logical thinking is in no way connected with it. For the individuals mentioned above, mathematics is quite meaningless, and only needless torment. The truth is that mathematics presupposes a definite type of psychological make-up that is by no means universal and that cannot be acquired. For those who do not possess this ability mathematics becomes merely a subject to be memorized, just as one memorizes a series of meaningless words. Such persons may, however, be highly gifted in every other way, and may either possess already the capacity for logical thinking, or have a better chance of acquiring it by direct instruction in logic. Strictly speaking, of course, a deficiency in mathematical capacity is not to be taken as an individual peculiarity. However, it serves to show in what way a curriculum may sin against the psychological peculiarity of the pupil.[3]

Second, the feeling for beauty in mathematics is infectious. It is caught, not taught. It affects those with a flair for the subject. I well remember when it happened to me, as a very young undergraduate of Bristol University. It was a seminal experience in life.

The late Peter Frazer, Lecturer in Mathematics, a lovable man and a brilliant teacher, was discussing cross ratios with a mathematics set. Swiftly he chalked on the blackboard a fan of four straight lines, crossed them with a transversal and wrote a short equation; he stepped down from the dais and surveyed the figure. I cannot of course recall precisely what he said but it went something like this. Striding rapidly up and down between the class and the blackboard, waving his arms about excitedly, with his tattered gown, green with age, billowing out behind him, he spoke in staccato phrases: "Och, a truly beautiful theorem! Beautiful! ... Beautiful! Look at it! *Look at it!* What simplicity! What economy! Just four lines and one transversal." His voice rises in a crescendo. "What elegance! *Any* lines, *any* transversal! Its generality is *astonishing*." Then, muttering to himself "Beautiful! ... beautiful! ...," he stopped, slightly embarrassed (he was from Aberdeen), and returned to earth.

The students were amused. But not all. Sparks from that blazing enthusiasm fell on at least one boy. He took fire and that fire was never extinguished. Hence this book.

SPIRITUAL VALUES

I must not close this introductory chapter without bearing witness to the *spiritual values* accruing from the pursuit of mathematics from the aesthetic standpoint. I have stressed the delight that derives from beholding beauty. But richer blessings may reward the sincere student. This is the testimony of many who have been disciplined by close association with the study of Nature and her interpreter, mathematics. Their testimony is that these disciplines can provide comfort, charm, edification, delight and blessings. I will attempt to justify these five nouns by quotations from five writers whose names stand out on the scroll of fame.

1. COMFORT: Johannes Kepler.

If there is anything that can bind the heavenly mind of man to this dreary exile of our earthly home and can reconcile us with our fate so that one can enjoy living,—then it is verily the enjoyment of the mathematical sciences and astronomy.

2. CHARM: Lord Rayleigh.

Some proofs command assent. Others woo and charm the intellect. They evoke delight and an overpowering desire to say "Amen, Amen."

3. EDIFICATION: Morris Kline.

Perhaps the best reason for regarding mathematics as an art is not so much that it affords an outlet for creative activity as that it provides spiritual values. It puts man in touch with the highest aspirations and loftiest goals. It offers intellectual delight and the exaltation of resolving the mysteries of the universe.[4]

4. DELIGHT: Tagore.

Somewhere in the arrangement of this world there seems to be a great concern about giving us delight, which shows that, in the universe, over and above the meaning of matter and forces, there is a message conveyed through the magic touch of personality.... Is it merely because the rose is round and pink that it gives me more satisfaction than the gold which could buy me the necessities of life, or any number of slaves.... Somehow we feel that through a rose the language of love reached our hearts.[5]

5. BLESSINGS: Wordsworth ("Lines Written above Tintern Abbey").

...that Nature never did betray
The heart that loved her; 'tis her privilege
Through all the years of this our life, to lead
From joy to joy: for she can so inform
The mind that is within us, so impress
With quietness and beauty, and so feed
With lofty thoughts, that neither evil tongues,

Rash judgments, nor the sneers of selfish men,
Nor greetings where no kindness is, nor all
The dreary intercourse of daily life,
Shall e'er prevail against us, or disturb
Our cheerful faith, that all which we behold
is full of blessings....

PHILOSOPHY

Finally, I hope the following pages may encourage the reader who takes mathematics seriously to cultivate a philosophical attitude to the subject; and not only about beauty in mathematics but also beauty in a wider context, to ask such questions as: What is beauty? What is the status of the aesthetic faculty? Has it a practical value? Whence is it? Has it served a purpose in human evolution? What are its long-term prospects?

It is difficult to confine beauty to either objective or subjective categories. It seems to be more satisfactory to regard it as an interaction between the mind and an object or an idea which arouses emotion. It would follow that the discovery of beauty either in the world of nature or in mathematics is indicative of some feature in the structure of the mind. For example, the impossibility of conceiving a finite universe, or a straight line that, however long, cannot be produced, points ineluctably to a constituent element in the mental fabric as much as to a feature of the universe. Again, the non-existence of an ultimate particle—one that cannot be subdivided—is not so much a fact of atomic physics as an inexorable mental necessity. The difficulty of conceiving an undifferentiated continuum or of imagining "action at a distance" are other examples of mental limitations.

These considerations provide a clue to indicate in which general direction we might look to find an explanation of the source of aesthetic pleasure. We should seek to discover what we can about the anatomy of the human psyche, which has been slowly evolved in parallel with the development of man's physical frame over a period of hundreds of thousands of years. One might accordingly anticipate that mental elements of great antiquity might be relevant to those types of beauty of which the appreciation is common to the whole human race, e.g., color contrast, rhythm, form, and others of the kind.

Carl G. Jung speculated as follows in his book, *Man and his Symbols*:

> Just as the ~~human body~~ represents a whole museum of organs, each with a long ~~evolutionary history~~ behind it, so we should expect to find that the mind is organized in a similar way. It can no more be a product without history than is the body in which it exists.... I am referring to the biological, prehistoric, and unconscious development of the mind in archaic man, whose psyche was still close to that of the animal.
>
> This immensely old psyche forms the basis of our mind, just as much as the structure of our body is based on the general anatomical pattern of the mammal. The trained eye of the anatomist finds many traces of the original pattern in our bodies. The experienced investigator of the mind can similarly see the analogies between the dream pictures of modern man and the products of the primitive mind, its "collective images" and its mythological motifs.

As an example of this one may point to the fact that one of the secrets of effective poetry is its power to bring to the surface mind (i.e., the conscious mind) primordial images which are deeply buried constituents of the racial unconscious common to all mankind. The poet who conjures up the archetypes "speaks with a thousand tongues."

A more relevant example might be the visual beauty of a sine curve or the comparable aural beauty of rhythm. Among the most ancient experiences of man—indeed, of his mammalian ancestors —is the inescapable association of rhythmic motion in the womb with the euphoria generated during nine months of pre-natal existence. It is the interaction between this (subjective) age-old, "fossilized" feature of mental structure and the (objective) sight of a sine curve or sound of rhythm in music which we call beauty.

This is perhaps sufficient to show that the serious study of mathematics can produce ~~philosophic speculation~~ as well as aesthetic appreciation. That it may also become a "chief source of happiness" is borne out by Bertrand Russell, who began Euclid at eleven, with his eighteen-year-old brother as tutor. In his *Autobiography* he testifies:

> This was one of the great events of my life, as dazzling as first love. I had not imagined there was anything so delicious in the world.... From

that moment until ... I was thirty eight, mathematics was my chief interest and my chief source of happiness.

It is the writer's hope that the reader with a humbler mathematical gift than Russell's may find the following pages a "source of happiness."

CHAPTER I

The Texture of Beauty

Before launching out on our main topic, beauty in mathematics, it will be worth while to convince ourselves that the effort required to learn to appreciate aesthetic values is justified by the pleasure it offers. The conviction is born of experience; and we shall soon discover that it is shared by many of the world's wisest men. An example of ancient standing, written before the Christian era (*Ecclesiasticus* 43, vv. 11–12), magnifies one of the most familiar of beautiful objects:

Look upon the rainbow, and praise him that made it; very beautiful it is in the brightness thereof. It compasseth the heaven about with a glorious circle, and the hands of the most High have bended it.

A further quotation relating to the same example of beauty will serve to underline one of the important lessons of these chapters. For aesthetic appreciation there are two requirements: the first is *given*, the second *acquired*. The first is from nature—by inheritance; the second from nurture—by education.

If the poet sees beauty in a rainbow—

My heart leaps up when I behold
A rainbow in the sky...

—so does the physicist in the laws governing its manifestation:

His heart leaps up, too, as he discovers how the light of day is reflected, chromatically refracted, reflected again and dispersed by gently falling water spheres into a thousand hues, conforming the while to lovely theorems of mathematics so simple in some aspects that the schoolboy may understand, so complex in others as to defy analysis. [1]

The surface beauty of the rainbow—"very beautiful it is in the brightness thereof"—is appreciated by all men: it is *given*. But the buried beauty, uncovered by the industrious researches of the physicist, is understood only by the scientifically literate. It is *acquired*: education is essential.

DEFINITIONS

It is difficult to define beauty, as we shall see; but there is much impressive testimony to the importance of the emotions that beauty calls forth. Mohammed said:

If I had only two loaves of bread, I would barter one to nourish my soul.

A more modern witness, Richard Jefferies, wrote:

The hours when we are absorbed by beauty are the only hours when we really live.... These are the only hours that absorb the soul and fill it with beauty. This is real life, and all else is illusion, or mere endurance.

Beauty is a word which has defied the efforts of philosophers to define in a way that commands general agreement. Yet it does not need a philosopher's wisdom to utter a few meaningful words about it. One incontrovertible statement might be: beauty arouses emotion. This, being sufficiently indefinite, needs no qualification. The case would be different if we said, "This lovely artifact always arouses pleasurable emotion in everyone who sees it," for we know that some people, confronted by beauty which moves others, are entirely unresponsive to it. This appears to justify the familiar aphorism: "Beauty lies in the eye of the beholder." Whether this be true or not, it is certain that no philosopher, however erudite, can contradict me when I say in sincerity, concerning some experience,

"For me, that is beautiful." But whether beauty is subjective or objective or both is an unresolved metaphysical problem.

According to the *Shorter Oxford English Dictionary* beauty is:

that quality or combination of qualities which affords keen pleasure to the senses, especially that of sight, or which charms the intellectual or moral faculties.

Only a part of this wide definition concerns us here. We are not interested at the moment in, for example, "the beauty of holiness" which "charms" the moral faculties. Our interest lies in the combination of qualities which charms the intellect. "Combination of qualities" reminds us that the experience of beauty is not a simple, but a complex experience. In mathematics it may be compounded of surprise, wonder, awe, or of realised expectation, resolved perplexity, a sense of unplumbed depths and mystery; or of economy of the means to an impressive result. When the mathematician refers to beauty in mathematics, we infer that he has had experience of some or all of these qualities.

Before we turn to the consideration of particular types of beauty, it is profitable to think of it in a wider context.

EVOLUTION OF AESTHETIC FACULTY

Taking a teleological view-point, we might begin by asking whether the universal human thirst for beauty serves a useful purpose. Physical hunger and thirst ensure our bodily survival. The sex drive takes care of the survival of the race. Fear has survival value. But—to put the question crudely—what is beauty for? What personal or evolutionary end is met by the appreciation of a rainbow, a flower or a symphony? At first sight, none. Why, if I have two loaves, should I "sell one and buy a lily"? Many of our appetites have been developed in the course of human evolution for a utilitarian purpose in the material environment of our mundane existence. Does this suggest a realm of another natural order? That it points to a definitive view of the nature of the human psyche is a conclusion which seems unavoidable. Before we develop this, let us remind ourselves of how some philosophers, both ancient and modern, have regarded beauty.

Plato, in the *Symposium*, has much to say about progress from aesthetic appreciation to the enjoyment of "absolute beauty." He recounts an inspired speech by Socrates in a dramatic dialogue at the "Dinner Party." Socrates modestly attributes his views to his "instructress"—a woman from Mantinea, called Diotima. The following excerpts are relevant to our subject:

The man who would apply himself to this goal must begin, when he is young, by applying himself to the contemplation of physical beauty.... The next stage is for him to reckon beauty of soul more valuable than beauty of body....From morals he must be directed to the sciences and contemplate their beauty also....[The man] who has directed his thoughts towards examples of beauty in due and orderly succession will suddenly have revealed to him as he approaches the end of his initiation a beauty whose nature is marvellous indeed, the final goal, Socrates, of all his previous efforts. This beauty is first of all external; it neither comes into being nor passes away; next, it is not beautiful in part and ugly in part, nor beautiful at one time and ugly at another....He will see it as absolute, existing alone with itself, unique, external, and all other beautiful things as partaking of it....

This above all others, my dear Socrates, (the woman from Mantinea continued) is the region where a man's life should be spent, in the contemplation of absolute beauty. Once you have seen that, you will not value it in terms of gold or rich clothing or the beauty of boys and young men.... What may we suppose to be the felicity of the man who sees absolute beauty in its essence, pure and unalloyed, who, instead of a beauty tainted by human flesh and colour and a mass of perishable rubbish, is able to apprehend divine beauty where it exists apart and alone? Do you think that it will be a poor life that a man leads who has his gaze fixed in that direction, who contemplates absolute beauty with the appropriate faculty and is in constant union with it?

Turning from an ancient to a modern philosopher, we may consider the views of the Italian philosopher, Benedetto Croce. His position is that beauty is an attribute of that which expresses feeling. Music, as Plato recognized, expresses human emotion very vividly; it is *lento, vivace, con brio*, etc. Beauty is seen in colors that are gay or somber. And there is the beauty of scenery:

> Bright robes of gold the fields adorn,
> The hills with joy are ringing,
> The valleys stand so thick with corn
> That even they are singing.

Clouds are lonely or angry; the morn is smiling; the oak is majestic; a mathematical theorem is elegant, its proof neat.

Wordsworth said of poetry that it was "emotion recollected in tranquillity."

Following Croce, then, we may take it that the aesthetic experience supervenes when some material or mental entity, to which for that reason we attribute "beauty," stimulates pleasurable emotion.

Now emotions are regarded by psychologists as activities of the unconscious mind, so that the aesthetic experience is the resuscitation of subliminal emotions, and beauty is the power to evoke these emotions. This takes us into deep waters and we will postpone a discussion of the function of the unconscious in mathematical studies to a later chapter.

UNITY IN VARIETY

For a modern view of the nature of beauty we may turn again to J. Bronowski:

> When Coleridge tried to define beauty, he returned always to one deep thought; beauty, he said, is unity in variety! Science is nothing else than the search to discover unity in the wild variety of nature,—or, more exactly, in the variety of our experience. Poetry, painting, the arts are the same search, in Coleridge's phrase, for unity in variety.[2]

A WORKING HYPOTHESIS

A student who aspires to gain an insight into a philosophy of the beauty that is latent in mathematics should fortify himself with some form of working hypothesis concerning beauty in a wider context, with the help of modern views on the nature of the human psyche as developed by Sigmund Freud, Carl Jung and other psychologists. Such an hypothesis will not have the status of a theory. While it will, of course, require modification and amplification as new knowledge is gained, it is not thereby invalidated as a viable frame of reference, of which the function is to maintain a logical sequence among numerous data.

Fundamental to such an hypothesis is the recognition that the aesthetic experience is an emotional, rather than a rational mental

activity. Merely to state this basic fact is to realize that we shall
not make much progress in understanding without admitting the
relevance of what has been called "the greatest discovery of the
nineteenth century"—the *subconscious mind.* Though psychologists have found this topic a fertile source of differences of opinion,
they are agreed concerning its importance in interpreting mental
activity. It is invoked to explain such phenomena as hypnotic
trance, dreams, narcosis, dual personality, mental disorders and
much more. Its value for our present purpose is that it provides a
clue to the understanding of aesthetic feeling.

STRUCTURE OF THE PSYCHE

Psychologists often use the wider terms *psyche* and *psychic* in
place of mind and mental, which are normally applied to the
conscious mind only. Psychic activity is no less real than physiological activity: the psyche has its own structure and is governed
by its own laws. A montage of the psyche as seen by the pioneer
psychoanalyst Carl Jung would cover four main concepts:

1. The *conscious mind,* or "surface" mind, the seat of conscious
mental activity.

2. The *preconscious,* sometimes pictured as forming the periphery of the conscious mind, sometimes as the stratum below the
surface mind. Our memories of recent events, now removed from
the focus of attention, are stored herein. From here they may be
voluntarily recalled—recollected.

3. The *subconscious or personal unconscious.* The conscious
mind, according to Jung, is

based upon and results from an unconscious psyche which is prior to
consciousness, and continues to function together with, or despite
consciousness.[3]

The dipstick of introspection cannot plumb this layer of the
psyche. Unconscious activity is only exceptionally recognized by
the individual, despite the fact that unconscious motivation is one
of the prime facts of life. In the unconscious are stored countless
forgotten memories which, while they cannot be recalled at will,
are nevertheless made manifest in dreams, in hypnotic trance and
through other means.

4. The *collective unconscious*, according to Jung, forms a lower stratum of the psyche than the personal unconscious. It is the source of instinctive behaviour, an instinct being defined as "an impulse to action without conscious motivation." Instinctive behavior is inherited: it is determined by the history of the race. So are what Jung calls "primordial images" or "archetypes," which were formed at low mental levels during the tens of thousands of years of the evolutionary history of primitive man, our remote human ancestors, by the constant recurrence of universal emotional experiences common to all, e.g., the alternation of day and night, seasonal changes, hunger and thirst, flight from danger, the mountains and the oceans, storm and tempest, the sanctuary of hearth and home.

As a mnemonic, the structure of the psyche may be compared to an ocean island. The land above the water surface represents the conscious mind, the area uncovered at low tide depicts the preconscious; the vast, hidden mass of rock below the ocean represents the unconscious which rises from an ocean bed standing for the collective unconscious. The scientists halt here, but the theologians (notably Tillich) speak of the deepest level of all, which undergirds the ocean bed, as "the ground of our being," and equate it with God.

EMOTIONAL ACTIVITY

If we may assume that the evolution of psychic potentialities through geological ages has run parallel to the development of the nervous system and the brain, it would appear that, historically, emotional life which we share with the higher animals must precede intellectual development and be associated with the primitive parts of the nervous system. Incidentally, this also controls the visceral activities of the body and that is why a public performer afflicted with "nerves" sometimes has cause to observe the connection between his emotion and the activity of his intestines! We must accordingly conclude that the personal unconscious, as well as the collective unconscious, is the arena of the emotions as well as the storehouse of emotive memory complexes.

Now, since the structure of the nervous system is inherited, it is not unreasonable to suppose that the physiological conditions favorable to the animation of primordial emotions of the collective unconscious are also handed down from generation to generation. It is accordingly natural to postulate that the tendency of the psyche to make certain broad aesthetic judgments relating to the common human environment is inherited. H. J. Eysenck refers to the hypothesis, based on experiments, that

there exists some property of the nervous system which determines aesthetic judgments, a property which is biologically derived.... One deduction, for instance, might be that this ability (aesthetic judgment) should be very strongly determined by heredity; there is already some evidence for this point of view....[4]

AESTHETICS

Let us now turn from general considerations to the particular case of the emotion generated by the interaction between an object of beauty and an observer—the aesthetic feeling. If the foregoing sketch of a working hypothesis is on the right lines, then the aesthetic experience consists in the levitation from the unconscious to the surface mind of a memory complex activated by an association mechanism sequential to the visual or aural contemplation of the beautiful object. It is not difficult to guess the nature of these hidden memory complexes: they arise from the immemorial terrestrial environment of man. The complexity of this defies analysis, but it will make our meaning clear if we point to a few specimen experiences which have been familiar to both men and animals for a million years: (i) color contrasts, (ii) the gravitational field, (iii) bird song, human conversation and vocal music.

i. Our pleasure in color is shared with some of the vertebrates. Dr. W. H. Thorpe, describing the Bower birds of Australia and New Guinea, states that they build bowers for courtship with

brightly coloured fruits or flowers which are not eaten but left for display and replaced when they wither.... They stick to a particular colour scheme. Thus, a bird using blue flowers will throw away a yellow flower inserted by the experimenter, while a bird using yellow flowers will not tolerate a blue one.[5]

Dr. Thorpe quotes Robert Bridges: "Verily it may well be that sense of beauty came to those primitiv bipeds earlier than to man." In that case we should not be surprised if mankind's collective unconscious, carrying such an inheritance from the lower creation, is deeply stirred at the sight of flowers, a colorful sunset or a rainbow, rousing the conscious mind, in due sequence, to the aesthetic response.

ii. Apart from its colors, the gentle curve of the rainbow would, through association, stimulate memories stored in the collective unconscious by the ever-present phenomena of the earth's gravitational field—the lovely parabolic path of a flying stone, or spear or arrow, of the water drops of a fountain or cascade. All parabolas have the same unique shape, of which the mighty circular arc of the rainbow is reminiscent.

iii. Similar considerations apply to the beauty of music. Dr. Thorpe remarks that

...it is perhaps plausible that the intervals which are acceptable to the human ear, as normal and natural for music, are in fact those intervals which were first offered to the ancestors of man by bird song. Other animals do not have much in the way of song; but the fundamental intervals of human and bird song are the same; and highly developed bird song was audible at man's first appearance in time. Since man always had bird song all around, impinging on his ears, is it not reasonable to suppose that he developed a musical signal system by imitating the birds?[6]

MUSIC

Music is the language of the unconscious mind par excellence. As we shall argue in chapter VI, primordial racial memories are brought to the surface more readily by music than by natural scenery or any other art; it seems to be possible to relate familiar features of music to archaic experiences of humanity.

It is music that provides the strongest support for our thesis that aesthetic experience consists in the interaction between the universal primordial images buried in the unconscious and an external artifact or natural object which we call beautiful.

The incomparable power of music to move a listener to the depths of his being is well-known; it will, on occasion, bring him to tears. What is the explanation of the power of this stimulus which is unparalleled in the other arts? If our thesis is tenable it must be that music is for some reason an unusually effective agent for bringing to the surface archaic images and memories stored in the unconscious. As Jung remarks (see p. 77), "The man who speaks with primordial images speaks with a thousand tongues.... That is the secret of effective art." Now musical expression can stimulate archaic emotional experiences very effectively—fear by *agitato*, mourning by *molto legato*, excitement by *prestissimo*, sanctuary by *rallentando* succeeded by the tonic or *home* note, and in similar ways. These universal emotionally charged experiences become effective when they are raised from the deep unconscious to the surface mind, and it happens that music, unlike any of the other arts, provides precise and powerful means of effecting this transfer.

When a hypnotist induces a trance in a suitable subject, he finds a simple, potent method in making rhythmic passes with his hands in front of the subject's eyes. This induces a light trance and increases the subject's suggestibility. He will then be ready to accept almost any suggestion—even an absurd one—that the hypnotist makes. In particular, as Eysenck points out,

> Under hypnosis a person can remember things which in the normal state he would be quite unable to recall. In fact, it is suggested that under hypnosis a person can be "regressed" to an earlier age and that in this condition he will experience again the events which were happening at that time and the emotions which they evoked in him.[7]

In the case of music, the rhythmic movements of the hypnotist's hands are replaced by those of the conductor's baton which is amplified by the rhythm of the music. When the "beat" is very strong, as in dance music among primitive tribes, the hearers become entranced. Even under the more familiar conditions of an audience listening to Western music, it may be presumed that most if not all listeners are under the influence of a very light trance which increases their suggestibility and facilitates the levitation of buried archaic images to the conscious mind.

It is along such general lines that we may look for the "secret of the effective art" of music and an explanation of the aesthetic pleasure resulting from it.

And now let us return to the consideration of the question of whether beauty serves a purpose in the scheme of creation. We have already seen that it appears to serve no utilitarian end. Many of our instincts and associated emotions have been evolved to ensure our bodily survival, but the emotion aroused by a physical object such as a cloud or a flower, or by a mental image such as an elegant mathematical theorem, has no such objective. The answer to the question which we posed in crude terms: "What is beauty for?" appears to be elusive. So much is this so that one is inclined to doubt whether it has any purpose and to dismiss the matter by asking impatiently, "Must all things have a *raison d'être*? Is not a thing of beauty a joy for ever, and, so far from being a means to an end, an end in itself?"

AESTHETIC PLEASURE UNIVERSAL

It seems to me to be important that we should have clear ideas in reference to this question, and I hope the reader, before he proceeds to the chapters which follow, will give careful consideration to the point of view now to be described. At first glance it may appear that the contemplation and appreciation of the beauty of, say, a mathematical theorem is an unimportant, even trivial, activity. On the contrary, it is, properly regarded, one of great significance. It would seem to be unlikely, *a priori*, that the whole human race should be endowed with the faculty to enjoy beauty unless it achieved some noble consummation. "Earth's crammed with heaven and every common bush aflame with God" to some purpose, surely? The power to appreciate beauty appears to be a human endowment and this suggests that we should seek its origin and its purpose in human nature—in that nature which distinguishes us from the animal creation. Thus, for an answer to our question, we are driven back to the explanation of our human nature given in *Genesis* 1, v. 26:

And God said, Let us make man in our image, after our likeness.

Here, I suggest, is the clue. Man is by nature a *creator*. After the likeness of his Maker, man is born to create: to fashion beauty, to originate new values. That is his supreme vocation. This truth awakens a resonant response deep within us, for we know that one of the most intense joys that the soul of man can experience is that of creative activity. Ask the artist. Ask the poet. Ask the scientist. Ask the inventor or my neighbor who grows prize roses. They all know the deep spiritual satisfaction associated with the moment of orgasm of creation.

CREATIVE ACTIVITY: EMPATHY

This deep joy has been thought by some to be the principal aim of education—more, the chief end of human life. In *The Education of the Whole Man*, L. P. Jacks writes:

What then is the vocation of the whole man? So far as I can make out, his vocation is to be a creator: and if you ask me, Creator of what?, I answer—creator of real values.... And if you ask me what motive can be appealed to, what driving power can be relied on, to bring out the creative element in men and women, there is only one answer I can give; but I give it without hesitation—the love of beauty, innate in everybody, but suppressed, smothered, thwarted in most of us.... [8]

This inborn love of beauty, our human heritage, *must find* expression if we are to be happy. If the hunger for beauty remains unsatisfied, the effects are seen in loss of physical and mental health, so deep is the need.

We now approach the final stage in the argument of this chapter. It underlines a truth which it is important that all students of mathematics should understand, but (it is to be feared) very few do. If it could be expressed in one word, that word would be *empathy*. The German equivalent is *Einfühlung*—"feeling into."

We have spoken of a common experience—the joy associated with any form of creative activity, which a man has as a consequence of his having been made in the image of his Creator. And we have interpreted the mystery of the nature and purpose of beauty by recalling the familiar fact that the inborn faculty of aesthetic appreciation constitutes the motive for the creation of objects of beauty. And now we have to meet the natural objection that many would raise: they have had no experience of creative activity. They have added nothing to the store of beauty, their own ideas have been neither new nor original. They have never known the luminous moment of inspiration which widened the bounds of knowledge. They can appreciate, but cannot create beauty.

The answer to this objection can be stated briefly. The act of creation and the act of appreciation of beauty are not, in essence, distinguishable. This is true whether the lovely object is a work of art, a musical composition or a mathematical theorem. In the actual moment of appreciation ("*I see! Yes, indeed I see! How beautiful!*"), the beholder experiences those precise emotions which passed through the mind of the creator in his moment of creation. With the help of the artist he himself shares the joy of creation. This important fact has been expressed with characteristic clarity by J. Bronowski:

> The discoveries of science, the works of art are explorations—more, are explosions, of a hidden likeness. The discoverer or the artist presents in them two aspects of nature and fuses them into one. This is the act of creation in which an original thought is born, and it is the same act in original science and original art.... [This view] alone gives a meaning to the act of appreciation; for the appreciator must see the movement, wake to the echo which was started in the creation of the work. In the moment of appreciation we live again the moment when the creator saw and held the hidden likeness.... We re-enact the creative act, and we ourselves make the discovery again.... The great poem and the deep theorem are new to every reader, and yet are his own experiences, because he himself re-creates them. They are the marks of unity in variety, and in the instant when the mind seizes this for itself, the heart misses a beat.[9]

This passage, which illuminates the meaning of *empathy*, should be understood by all who seek the aesthetic experience. In particular, the reader of the following pages, whether his interest is focused on the golden cuboid, or the dodecahedron, or the logarithmic spiral or the genealogy of the drone bee, should realize that, in the act of appreciation, he is re-enacting the creative act and, attracted by beauty, is experiencing himself the joy of creative activity. He is in fact, in Kepler's phrase, "thinking God's thoughts after Him."

CHAPTER II

The Divine Proportion

Geometry has two great treasures: one is the theorem of Pythagoras; the other, the division of a line into extreme and mean ratio. The first we may compare to a measure of gold; the second we may name a precious jewel.

KEPLER [1571–1630]

Some of the earliest references to the pleasures of mathematics are linked with the name of the Greek philosopher, Pythagoras (569–500 B.C.), who observed certain patterns and number relationships occurring in Nature. Whether or not he speculated concerning the simpler facts of crystallography, as some writers declare, it is certain that he studied, and was very interested in, the dependence on its length of the pitch of the note emitted by a vibrating string. In particular, he noted the curious fact that the lengths which emitted a tonic, its fifth, and its octave, were in the ratio 2:3:4. The explanation of the order and harmony of Nature was, for Pythagoras, to be found in the science of numbers. He speculated that harmonious sounds were emitted by the heavenly bodies as they described their celestial orbits; this is the "harmony of the spheres," a notion which Shakespeare found congenial (*Merchant of Venice*, V, i, 57):

> There's not the smallest orb which thou behold'st,
> But in his motion like an angel sings,
> Still quiring to the young-eyed cherubins.

It is noteworthy that one of the most harmonious combinations of notes, the major triad, has relative frequencies expressed in ratios of small integers, *viz.*, 4:5:6. An explanation of this curious observation was advanced by H. L. F. von Helmholtz (1821–1894) on the basis of the presence or absence of "beats" between the overtones of these notes.

We shall see later (Chap. VI) that pleasure in mathematics is sometimes related to the appreciation of music. The experiments of Pythagoras are relevant to this, and it may turn out that there is a connection between the major sixth and the golden section.

THE GOLDEN SECTION

Another fact known to the school of Pythagoras was that there are five, and only five, regular convex solids, each of which could be circumscribed by a sphere: the tetrahedron, cube, octahedron,

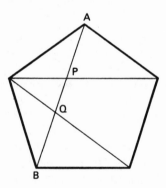

Fig. 2.1. Pentagon: diagonals

icosahedron and dodecahedron. A taste for "the mysteries" led the ancient Greek to ascribe a special significance to the last-named of these: its twelve regular facets corresponded to the twelve signs of the zodiac. It was a symbol of the universe. Moreover, each pentagonal face, being associated with the golden section, had a special interest for the Pythagoreans. The point of intersection P of two diagonals divides each in the golden ratio (Fig. 2.1). P divides AQ and AB internally and QB externally in this ratio. Another fact within the knowledge of these ancient geometers was that the ratio of the radius of the circumcircle of a regular decagon to a side is the golden ratio.

The problem of finding the golden section of a straight line is solved in Euclid II, 11. It has therefore been a topic of interest to mathematicians for more than twenty centuries.

Let a line AB of length l be divided into two segments by the point C (Fig. 2.2). Let the lengths of AC and CB be a and b

Fig. 2.2. Golden cut

respectively. If C is a point such that $l:a$ as $a:b$, C is the "golden cut" or the golden section of AB.

The ratio l/a or a/b is called the golden ratio. In the terminology of the early mathematicians AB is divided by C in "extreme and mean ratio." Kepler called it "the divine proportion."

There seems to be no doubt that Greek architects and sculptors incorporated this ratio in their artifacts. Phidias, a famous Greek sculptor, made use of it. The proportions of the Parthenon illustrate the point (see Chap. V, Fig. 5.2).

PHI

It was suggested in the early days of the present century that the Greek letter ϕ—the initial letter of Phidias's name—should be adopted to designate the golden ratio. The ubiquity of ϕ (*Phi*) in mathematics aroused the interest of many mathematicians in the Middle Ages and during the Renaissance. In 1509 there was published a dissertation by Luca Pacioli, *De Divina Proportione*, which was illustrated by Leonardo da Vinci. Reproduced in a handsome edition in 1956, it is a "fascinating compendium of *Phi*'s appearance in various plain and solid figures."[1] We shall in following chapters come across many examples of the appearance of *Phi* in unexpected places.

If, in figure 2.2, *BA* is produced to *D* where $AD = a$, then the following relationships hold:

Since $l/a = a/b$, $l \cdot b = a^2$, and from $(a/l) + 1 = (b/a) + 1$ we obtain $(l + a)/l = l/a$, i.e., $BD/BA = BA/AD$, so that A is the golden section of *BD*.

The numerical value of *Phi* is easily calculated. In figure 2.2 let $AC = x$, $CB = 1$, so that $AC/CB = x = Phi$.

$$\frac{x + 1}{x} = \frac{x}{1}, \qquad \text{i.e., } x^2 - x - 1 = 0.$$

The positive solution of this is $x = (1 + \sqrt{5})/2 = 1.61803$, giving the value of *Phi* to five decimal places. The positive solution we shall denote by ϕ, and the negative solution by ϕ'.

If, instead of $CB = 1$, we take $AC = 1$ and $CB = x'$, then

$$\frac{x' + 1}{1} = \frac{1}{x'}, \qquad \text{i.e., } \quad x'^2 + x' - 1 = 0$$

The positive solution of this is $x' = (\sqrt{5} - 1)/2 = 0.61803$. This, prefixed by the negative sign, we call ϕ'.

Thus, curiously, ϕ' turns out to be the negative reciprocal of ϕ; that is, $\phi \cdot \phi' = -1$. For

$$\frac{1}{\phi} = \frac{2}{1 + \sqrt{5}} = \frac{\sqrt{5} - 1}{2} = -\phi'$$

Phi is unique in this property: it is the only number, which, when diminished by unity, becomes its own reciprocal:

$$\phi - 1 = \frac{1}{\phi}, \qquad \text{i.e., } \quad \phi^2 - \phi - 1 = 0$$

Thus, ϕ and ϕ' are the solutions of $x^2 - x - 1 = 0$. We shall take ϕ to be the positive solution $(1 + \sqrt{5})/2$, and ϕ' to be the negative solution $(1 - \sqrt{5})/2$. It is evident from this, as it is from the properties of the roots of the quadratic equation, that

$$\phi + \phi' = 1 \quad \text{and} \quad \phi \cdot \phi' = -1$$

TO DIVIDE A STRAIGHT LINE IN THE GOLDEN SECTION

Let AB (Fig. 2.3) be the given straight line. Draw $BD = AB/2$ perpendicular to AB. Join AD.

With center D, radius DB, draw an arc cutting DA in E. With center A, radius AE, draw an arc cutting AB in C.

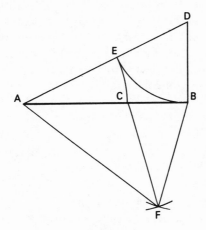

Fig. 2.3. Golden section: geometrical construction

Then C is the golden section of AB.

The proof that AC/CB is the golden ratio is left to the reader.

MULTIPLES OF THE ANGLE π/5 (36°)

The method described above of dividing a line in the golden section suggests a method of constructing an angle of 36° with ruler and compasses.

With center C (Fig. 2.3), radius CA, describe an arc. With the same radius, center B, cut the arc in F. Join A, B, C to F.

Then $\angle BAF = 36°$. Also $\angle CBF = 72°$ and $\angle ACF = 108°$.

The proof depends on showing that $FA/FB = CA/CB$, so that FC bisects $\angle AFB$.

It is then easily proved that $AF = \phi \cdot AC$.

THE PENTAGRAM STAR

The number of regular polygons which can be constructed in two-dimensional space is unlimited. The number of regular convex polyhedra in a space of three dimensions is five. How many regular four-dimensional figures are possible?

The Pythagoreans, who were interested in such matters, regarded the *dodecahedron* as being worthy of special respect. By extending the sides of one of its pentagonal faces to form a star, they arrived at the *pentagram*, or *triple triangle*, of figure 2.4, which they used as a symbol and badge of the Society of Pythagoras. By this sign they recognized a fellow member.

It is a rich source of golden ratios. The following 12 properties

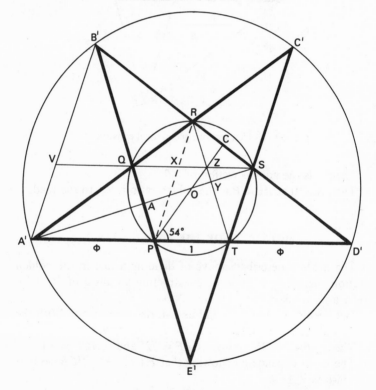

Fig. 2.4. Pentagram or triple triangle

are easily verified, taking R, r as the radii of the circumcircles of the pentagons $A'B'C'D'E'$ and P, Q, R, S, T respectively, and PQ as of unit length.

  i. $A'P = \phi$

  ii. $OA/r = \phi/2$

  iii. $OA'/r = \phi^2$

  iv. $OA'/OA = 2\phi$

  v. A diagonal such as QS has length ϕ.

  vi. If X is the point of intersection of two diagonals PR, QS, then

$$\frac{SX}{XQ} = \phi, \qquad \frac{PX}{XR} = \phi \quad \text{and} \quad \frac{B'X}{XT} = \phi$$

  vii. If SQ produced meets $A'B'$ in V, then, since VQS is parallel to $A'D'$,

$$\frac{B'V}{VA'} = \frac{B'Q}{QP} = \frac{B'X}{XT} = \frac{B'S}{SD'} = \phi$$

  viii. The lengths of the six segments $B'D'$, $B'S$, $B'R$, RS, RX, XZ are in geometric progression.

$$B'D' = \phi^3$$
$$B'S = \phi^2$$
$$B'R = \phi$$
$$RS = 1$$
$$RX = \phi^{-1}$$
$$XZ = \phi^{-2}$$

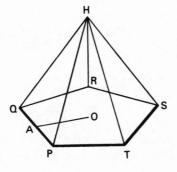

Fig. 2.5. Folded pentagram

The series is also an additive series: the sum of two consecutive members equals the next, e.g., $\phi + \phi^2 = \phi^3$.

ix. The length of a side of the pentagon $A'B'C'D'E'$ is ϕ^2.

x. $R/r = \phi^2$.

By folding $\triangle A'PQ$ about PQ and treating the other corresponding triangles similarly, so that A', B', C', D', E' meet in H, (Fig. 2.5) we have a pyramid of height OH.

xi. $OH/OA = 2$

xii. $OH/r = \phi$

THE PYTHAGOREAN BROTHERHOOD

The pentagram star was also regarded by members of the ancient society of Pythagoras as a symbol of health. Probably the five angles were denoted by the letters $\Upsilon\Gamma\Theta\Lambda$, the Greek word for health (Θ standing for the diphthong EI).

A Greek writer, Iamblichus, tells us that a member of the Pythagorean fellowship, while travelling far from home, stayed one night at a wayside inn. He fell ill, and despite the care of a sympathetic landlord, who tried at considerable expense to restore him to health, he died. Before his death, recognizing that his situation was desperate and being unable to compensate his host, he had obtained a board and on it had inscribed a pentagram. Giving this to the landlord he had requested that it might be fixed where all passers-by would see it. In due course a traveller riding by saw the symbol. Dismounting he made enquiries and, on hearing the story of the landlord, generously recompensed him. We may assume, I think, in view of his generous and disinterested treatment of a wayfaring student, that the landlord made no further use of the board inscribed with the triple star.

A rectangle, the sides of which are in the golden ratio, is called the golden rectangle. Its shape appears to have aesthetic attractions superior to that of other rectangles. The evidence, based on experiments in psychology, is presented in chapter V. Whatever the truth of the matter, there seems to be no doubt that Greek architects made use of this form in their designs. An example is seen in the representation of the Parthenon in chapter V (Fig. 5.2). More significantly, the golden rectangle is associated in a natural way with four of the five regular convex solids known to the Greeks.

THE FIVE PLATONIC SOLIDS

The Greeks took a mystical view of the five regular solids. This is hardly surprising for the forms are beautiful in themselves. No mathematical sophistication is needed for the appreciation of the charm of their outward appearance: that is the *given* element of their beauty. By contrast, that which is *acquired* by training and education demands considerable mental effort.

The five regular solids were treated by Euclid in Book XIII of the *Elements* but are associated with the name of Plato because of his efforts to relate them to the important entities of which he supposed the world to be made. The aura of mysticism with which the Greek geometers surrounded them persisted until the dawn in the sixteenth century of the scientific era. But the aesthetic appeal of what are still known as the Platonic solids is undiminished. Next to the writer's chair, for some years past, there has stood a small dodecahedron of white china!

We are now confronted with our first example of beauty in mathematics. The facts have been common knowledge among mathematicians for 2000 years. Much evidence of their aesthetic appeal in the past is on record.

The first point to note about the regular Platonic solids shown in figure 2.6 is that they are precisely five in number. A little thought shows that, while an infinite number of polygons may be drawn on a plane surface, it is not possible to construct more than five regular polyhedra in three-dimensional space. The surface of a regular polyhedron is bounded by congruent regular polygons. The simplest polygons that can form the surface are the equilateral triangle, the square and the pentagon. It is clear from figure 2.6 that we cannot form the corner of a polyhedron with fewer than three faces and that a corner may be formed by joining three, four, or five equilateral triangles. With six such triangles, the corner flattens into a plane. The same will happen if four squares are united. Similarly, three regular pentagons at a corner is a maximum. But hexagons, and polygons with more than five sides are all ruled out. This argument for the limitation to five regular

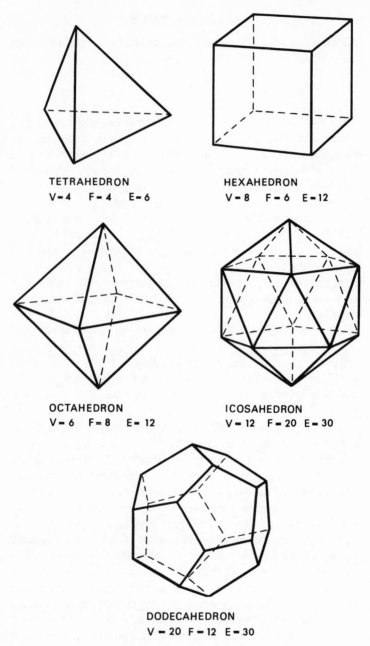

TETRAHEDRON
V = 4   F = 4   E = 6

HEXAHEDRON
V = 8   F = 6   E = 12

OCTAHEDRON
V = 6   F = 8   E = 12

ICOSAHEDRON
V = 12   F = 20   E = 30

DODECAHEDRON
V = 20   F = 12   E = 30

Fig. 2.6. The five Platonic solids

solids is the source of Euler's formula $V + F = E + 2$, where the letters stand for the number of vertices, faces, and edges respectively.

The second point of interest is that two pairs of the Platonic solids are reciprocal and the fifth is self-reciprocating in this sense: if the face centers of the cube are joined, an octahedron is formed, while the joins of the centroids of the octahedron surfaces form a cube. Similar relationship holds between the icosahedron and the dodecahedron. The join of the four centroids of the tetrahedron's faces makes another tetrahedron.

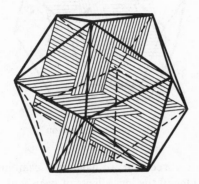

Fig. 2.7. Icosahedron

The third noteworthy feature is the relationship of the two pairs of reciprocal polyhedra to the golden rectangle (Figs. 2.7 and 2.8).

The Icosahedron. The twelve vertices of a regular icosahedron are divisible into three coplanar groups of four. These lie at the corners of three golden rectangles which are symmetrically situated with respect to each other, being mutually perpendicular, their one common point being the centroid of the icosahedron (Fig. 2.7).

The Octahedron. An icosahedron can be inscribed in an octahedron so that each vertex of the former divides an edge of the latter in the golden section.

The Dodecahedron. The centroids of the twelve pentagonal faces of a dodecahedron are divisible into three coplanar groups of four. These quadrads lie at the corners of three mutually perpendicular, symmetrically placed golden rectangles, their one common point being the centroid of the dodecahedron (Fig. 2.8).

The aesthetic appeal of the topics of this chapter cannot be doubted. Whether we can explain it or not, the fact that they have been enjoyed by sixty generations of men is good evidence.

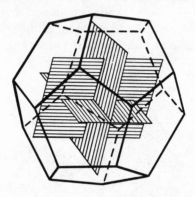

Fig. 2.8. Dodecahedron

It should be realized that we have only examined the surface of the subject. One of the ingredients of beauty in mathematics is its *depth*. Even in these familiar waters, "full many a gem of purest ray serene" awaits discovery by the explorer. There appears to be no limit, as Sir Edwin Arnold affirms in *The Light of Asia*:

> Shall any gazer see with mortal eyes
>   Or any searcher know with mortal mind—
> Veil after veil will lift—but there must be
>   Veil after veil behind.

CHAPTER III

Analysis of Beauty

When one is exploring unfamiliar country and, rounding a corner, gains suddenly a prospect of some well-remembered landmark; or when, engulfed in a crowd of strangers, one is suddenly confronted by the face of a friend, the reaction is often a feeling of surprised gratification. If the emotion evoked by the unexpected appearance of a familiar mathematical artifact is strong enough, we may feel that the artifact is "pleasing" or even "beautiful." Accordingly, surprise may be considered to be an occasional ingredient of mathematical beauty.

We may surmise that the sudden emergence from a mathematical process of a familiar concept or symbol in an unexpected relationship will sometimes evoke a faint pleasurable emotion, and, afflicted as we often are by the poverty of our vocabulary and incapable of describing our mental reaction more precisely, we may describe the novelty as "beautiful." An illustration of this may be useful.

Consider a sequence of integers formed according to the following rule:

$$u_{n-1} + u_n = u_{n+1}$$

Such a sequence is:

1, 3, 4, 7, 11, 18, $\cdots$, where $u_1 = 1$, $u_2 = 3$ and so on.

This is known as the Lucas sequence.

From this we obtain

$$u_{33}/u_{32} = 5781196/3570847 \tag{a}$$

Take at random any other sequence formed according to the same rule:

$$-3, +4, 1, 5, 6, 11, 17, 28, 45, \cdots$$

From this,

$$u_{25}/u_{24} = 160996/99501 \tag{b}$$

The first ratio (a) is $1.618\cdots$, but the second ratio (b) derived from a randomly chosen pair of initial terms is also $1.618\cdots$. In fact, for all values of n sufficiently large, any sequence formed according to the given rule produces the same results to three decimal places:

$$u_{n+1}/u_n = 1.618$$

SURPRISE, WONDER, CURIOSITY

The sense of surprise which this occasions is increased when it is realized that this ratio u_{n+1}/u_n approximates more and more closely to the golden ratio as n is increased, whatever may be the two initial terms. In fact,

$$\lim_{n \to \infty} u_{n+1}/u_n = \phi$$

A *pretty* result? What constitutes the essence of the aesthetic appeal of this outcome of simple mathematics? We seem to have no single word for it. It appears to be compounded of a mixture of archaic emotions. There is surprise at the unexpected encounter; there is also both curiosity and wonder—making three of the flavors included in the idea of beauty. *Curiosity:* because one craves to understand why *Phi*, which permeates the pentagram and is at home in Platonic polyhedra, should also be the limit of a ratio initiated so casually and generated as described above, a series which is apparently not even remotely related to the geometry of the Greeks. *Wonder:* because the conviction grows stronger that we have chanced on an unexplored world which, like the universe around us, appears to have no boundaries. There must, we speculate, be other discoveries to be made here by the inquiring mind. Suppose, for example, we chose negative numbers as the first two members of our sequence: $u_1 = -1$, $u_2 = -5$, $\cdots$, etc. We should find that dividing the greater by the lesser would give us (for example)

$$u_{11}/u_{12} = -304/-500 = 0.618$$

and so we should suspect that

$$\lim_{n \to \infty} u_n/u_{n+1} = \phi' \qquad \text{(i.e., } 1/\phi\text{)}$$

And we should be right.

The reader may care to discover for himself, by similar, simple numerical approximations, what would be the result of deducting unity from

$$\lim_{n \to \infty} u_{n+1}/u_{n-1}$$

STATUS OF PHI

What is the status of this small corner of the world of mathematics? Does it belong to the physical world? Are we to regard *Phi* as a "constant of nature"? Such constants appear often in the equations of the theoretical physicist. A familiar example is c, the velocity of electromagnetic radiation, such as light, in matter-free space. Another example is Planck's constant, h, useful in the study of the atom, its nucleus, and its radiation. A third example is e, the elementary electric charge. This is not of course the same as e, the base of natural logarithms, but is it of the same genus? is it of the same world? has it the same sort of reality?

We meet with π in trigonometry, in the geometry of a spherical raindrop, in probability theory, as the sum of an infinite series and in other situations. We meet with $i = \sqrt{-1}$ in complex numbers and in the theory of alternating electric currents. We come across the three combined in the remarkable equation

$$e^{i\pi} = -1$$

The question of mathematical reality is touched on by G. H. Hardy in his delightful essay *A Mathematician's Apology*. After admitting that neither mathematicians nor philosophers agree on the "nature of mathematical reality," whether it is mental, constructed by ourselves, or outside and independent of us, he proceeds:

A man who could give a convincing account of mathematical reality would have solved very many of the most difficult problems of metaphysics. If he could include physical reality in his account, he would have solved them all.

I should not wish to argue any of these questions here even if I were competent to do so, but I will state my own position dogmatically in order to avoid minor misapprehensions. I believe that mathematical reality lies outside us, that our function is to discover or *observe* it, and that the theorems which we prove, and which we describe grandiloquently as our "creations," are simply our notes of our observations. This view has been held, in one form or another, from Plato onwards . . . .[1]

MATHEMATICS: A LANGUAGE

No one would dispute that mathematics is a language. It is the language of the exact sciences. Its "words" are well defined. A serious "essay" expressed in mathematical symbols has a tang of poetry about it for the *cognoscenti*. The written language of mathematics has evolved in the course of time into an efficient shorthand. A wealth of ideas can be expressed in a very economical manner. An example has just been given:

$$e^{i\pi} = -1$$

and this shorthand is, of course, intelligible to students with the requisite training.

The following passage, culled from George Temple's broadcast on "The Nature and Charm of Mathematics," as published in *The Listener*, is relevant here:

There is one question on which mathematicians are sharply divided. It is the fundamental question as to what mathematics is. . . . I have maintained that mathematics is the language of physics. . . . Then what becomes of pure mathematics? . . . a language can be considered in at least two different ways, either in relation to the purpose which it serves as a medium for the expression of ideas or in relation to its internal structure. The study of linguistic relations as revealed in grammar, syntax, and in comparative philology is a vital and necessary element for the appreciation of any language. I venture to make the suggestion that pure mathematics is in fact the philological aspect of the language of physics.

There is a great deal to be said for this point of view if we consider the dominant characteristics of pure mathematics. These may be summarized as consistency, coherence, abstraction, and creativeness. It is almost a truism to say that the pure mathematician is not interested in the truth of his statements, but only in their internal consistency. It is as true that the practical physicist can be just a little impatient of the need for perfect consistency. . . .

UBIQUITY OF PHI

The status of *Phi* is not unlike that of π, since, as will be exemplified in the following pages, it not only crops up in Hardy's mathematical reality that "lies outside us," but it reveals itself in the world of nature, associated with phyllotaxis, with the patterns of florets in flowers of the composite family such as the sunflower, with the shape of the nautilus sea shell, and with other natural objects.

One may set out to attack a simple problem in pure mathematics, with no thought of the golden section, only to find that *Phi* fills an important role in the solution! We have already met an example in chapter II (Fig. 2.5), when we created a pyramid from a pentagram and (in Hardy's phrase) made a "note of our observation" of its height. This experiment (original, though unlikely to be new!) was rewarded by the discovery that the ratio of the height to the radius of the circumcircle of the base was *Phi*. A pleasant result, recalling a definition of beauty as "that which pleases in contemplation."

PHI IN TRIGONOMETRY

Similar, though more difficult, examples of the ubiquity of *Phi* are found in the inscribed triangle problem of chapter VII (p. 93) and in the tetrahedron problem of chapter VIII (p. 108).

As a simple example let us solve the equation:

$$\sin 2\theta = \cos 3\theta$$

Can you see *Phi* lurking in these innocent symbols? Since the sine of an angle is the cosine of its complement, $2\theta + 3\theta = \pi/2$ or $\theta = \pi/10$.

The equation may be reduced to

$$4 \sin^2 \theta + 2 \sin \theta - 1 = 0$$

Thus, $\sin 18° = \frac{1}{2}(\sqrt{5} - 1)/2$ or $-\frac{1}{2}(\sqrt{5} + 1)/2$. Taking the positive value,

$$\sin 18° = -\phi'/2, \quad \text{whence} \quad \cos 36° = 1 - 2 \sin^2 18° = \phi/2$$

THE GOLDEN TRIANGLE

These and similar results are collected in figure 3.1 and the table that follows:

Angle		$(2 \sin)^2$	$(2 \cos)^2$
$\pi/20$	9°	$2 - \sqrt{\phi + 2}$	$2 + \sqrt{\phi + 2}$
$\pi/10$	18°	$\phi' + 1$	$\phi + 2$
$3\pi/20$	27°	$2 - \sqrt{\phi' + 2}$	$2 + \sqrt{\phi' + 2}$
$\pi/5$	36°	$\phi' + 2$	$\phi + 1$
$\pi/4$	45°	$\phi + \phi'$	$\phi + \phi'$
$3\pi/10$	54°	$\phi + 1$	$\phi' + 2$
$7\pi/20$	63°	$2 + \sqrt{\phi' + 2}$	$2 - \sqrt{\phi' + 2}$
$2\pi/5$	72°	$\phi + 2$	$\phi' + 1$
$9\pi/20$	81°	$2 + \sqrt{\phi + 2}$	$2 - \sqrt{\phi + 2}$

The solutions of the equation $x^2 - x - 1 = 0$ have been given as

$$\phi = 1.61803 \qquad \phi + \phi' = 1 \quad \text{and} \quad \phi \cdot \phi' = -1$$

$$\phi' = -0.61803 \qquad \phi^2 = \phi + 1 = 2.61803$$

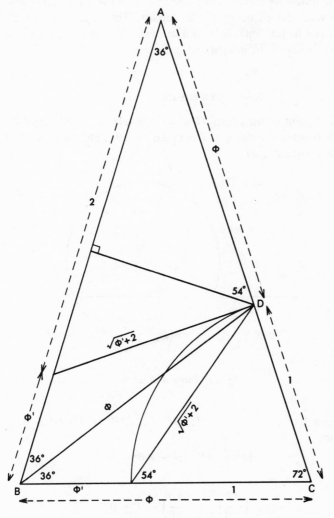

Fig. 3.1. Golden triangle

The following ratios are easily derived from figure 3.1:

$$\triangle ABC : \triangle ABD : \triangle DBC = \phi^2 : \phi : 1$$

◙ ◙ ◙

The following exercises in plane geometry are for students who have read the elements of the subject. They constitute simple additions to our anthology and are examples of the unexpected out-cropping of the golden section.

EXERCISE I

P is a point on the chord AB of a circle such that the tangent PT which touches the circle at T is equal to AB. Find the numerical value of the ratio $AP:AB$.

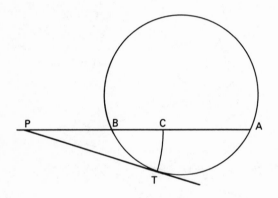

Fig. 3.2. Tangent problem

Using figure 3.2, $PT^2 = AP \cdot BP$, i.e., $AB^2 = AP(AP - AB)$, whence

$$AP^2 - AP \cdot AB - AB^2 = 0$$

or

$$\left(\frac{AP}{AB}\right)^2 - \left(\frac{AP}{AB}\right) - 1 = 0$$

Thus

$$\frac{AP}{AB} = \frac{1 + \sqrt{5}}{2} = \phi$$

If C is a point in PA such that $PC = PT$, find CA/CB.

We have $AP/AB = \phi$, so that

$$\frac{PB + BC + CA}{CA + BC} = \phi$$

But $PB = CA$. Hence

$$\phi = \frac{2CA + BC}{CA + BC} = \frac{2\alpha + 1}{\alpha + 1}$$

where

$$\frac{CA}{CB} = \alpha$$

Thus

$$\alpha = \frac{\phi - 1}{2 - \phi}$$

By substituting $(1 + \sqrt{5})/2$ for ϕ, we obtain $\alpha = \phi$. Hence

$$\frac{AP}{AB} = \frac{AB}{AC} = \frac{AC}{BC} = \phi$$

It follows that

$$\frac{AP}{AC} = \frac{AB}{BC} = \phi^2 \quad \text{and} \quad \frac{AP}{BC} = \phi^3$$

EXERCISE II—A FAMILIAR TRIANGLE

A well-known triangle of ancient fame was used by Egyptian surveyors for registering a right angle. A cord was divided by knots into three segments in the ratio $3:4:5$. When the ends are brought together to form a triangle, the angle subtending the 5-unit segment is a right angle.

Children learn this fact in school, but how many college students, even, realize that their old friend, the 3-4-5 triangle, is a hiding place for *Phi* and a few of its Fibonacci approximations?

Figure 3.3 shows a 3-4-5 triangle ABC. Using its labels, let the bisector of $\angle C$ meet AB in O. With center O, radius OB, describe a circle. It is easily seen that the hypotenuse CA is a tangent: let it touch the circle at B'. Join BB', let CO cut the circle in Q and

BB' in R. Let CO produced meet the circle in P. Then the following lengths of segments may be read off:

$$BC = 3, \qquad AB = 4, \qquad CA = 5, \qquad AB' = 2, \qquad B'C = 3$$

Because CO bisects $\angle C$, $AO/OB = AC/BC = \frac{5}{3}$; hence $AO = \frac{5}{2}$, $BO = \frac{3}{2}$. The line $CQROP$ is made up of the following segments:

$$PO = \tfrac{3}{2}, OR = \tfrac{3}{10}\sqrt{5}, RQ = \tfrac{3}{10}\sqrt{5}(\sqrt{5} - 1), QC = \tfrac{3}{2}(\sqrt{5} - 1)$$

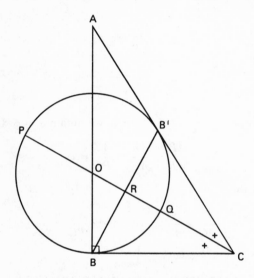

Fig. 3.3. The 3-4-5 triangle

Also:

$$CP = 3(1 + \sqrt{5})/2 = 3\phi$$
$$CQ = 3(\sqrt{5} - 1)/2 = -3\phi'$$

Successive Fibonacci approximations to *Phi* may be derived from the figure:

$$BC/BQ = \tfrac{2}{1}, \quad B'C/AB' = \tfrac{3}{2}, \quad AO/OB = \tfrac{5}{3}, \quad AB/BO = \tfrac{8}{5}$$

Also:

$$CP/PQ = \phi, \quad PQ/CQ = \phi, \quad OR/RQ = \phi/2$$

Thus Q divides CP in the golden section.

EXERCISE III—CONE PROBLEM

The circumference of the base of a right circular cone, of which the semi-vertical angle is 54° and the slant side measures one foot, is $\pi\phi$ ft., and the curved surface is $\frac{1}{2}\pi\phi$ sq. ft.

The proof of this, with the help of the table on p. 40, is left to the reader.

EXERCISE IV—THE FIVE DISCS PROBLEM

The following problem, solved by E. H. Neville,[2] makes an attractive addition to our anthology:

Five equal discs of unit radius are placed symmetrically as shown in figure 3.4 so that their centers form the corners of a

Fig. 3.4. The five disks problem

regular pentagon and their circumferences all pass through the pentagon's centroid O. What is the radius of the largest circular area covered by the five discs, i.e., What is the length of OA?

The answer is ϕ', the reciprocal of *Phi*!

Phi and Fi-Bonacci*

We have seen in the last chapter that *Phi*, conformably with its character of turning up unexpectedly in odd places, is connected with any sequence of integers formed according to the law that each term is the sum of the two preceding terms, whatever the first two terms may be: $u_{n+1} = u_n + u_{n-1}$. The ratio of successive terms, u_{n+1}/u_n, approximates more and more closely to *Phi* as n increases. We may take as a random example 5 and 2 as the initial terms, u_1 and u_2, giving the sequence

$$5, \ 2, \ 7, \ 9, \ 16, \ 25, \cdots, 280, 453, 733, \cdots, 13153, 21282, \cdots$$

from which we may determine approximations to *Phi*:

$$
\begin{aligned}
16/9 &= 1.7777\cdots \\
453/280 &= 1.6178\cdots \\
733/453 &= 1.6181\cdots \\
21282/13153 &= 1.61803\cdots
\end{aligned}
$$

This process brings us closer and closer to the value of *Phi*, which is $(1 + \sqrt{5})/2$. This value, to seven decimal places, is

* Filius Bonacci, son of Bonacci, shortened to Fibonacci.

1.6180340. A few calculations will show that the approximations oscillate, being alternately greater than and less than *Phi*:

$$453/280 = 1.6178\cdots < \phi, \qquad 733/453 = 1.6181\cdots > \phi.$$

In the absence of any restriction on the two initial members of the series, we may begin with the simplest; and this gives the Fibonacci series, so called by Edward Lucas in 1877:

$$0, 1, 1, 2, 3, 5, 8, 13, 21, \cdots$$

THREE METHODS OF CALCULATING

Using the Elliott 803 computer of Bath University, which reads the Algol computer language, the writer was able, by kind permission of the mathematics department, to prepare a program in the Algol language for calculating *Phi* to as many places as he required at the time. It is interesting to compare the speeds of the three methods of performing the calculation which are available today.

To calculate $u_{40}/u_{39} = 102,334,155/63,245,986$ by the handwriting methods available since the day of Fibonacci, and to check the result, would occupy a time measured in hours. To perform the same operation by the use of his "Britannic" desk calculator, the writer required a period measured in minutes. Using the Bath University computer he achieved the same end in a matter of less than 5 seconds.

An IBM 1401 computer operating in California[1] produced Fibonacci numbers 4000 digits in length. The article quoted gives the following interesting results:

TERM OF FIBONACCI SERIES	NO. OF DIGITS
u_{476}	100
u_{954}	200
u_{1433}	300
u_{1911}	400
. . . .	
u_{11003}	2300
u_{11004}	2300
. . . .	
u_{19137}	4000

The value of *Phi* was determined from the ratio u_{11004}/u_{11003}, each number containing 2300 digits, which are reproduced in full in the journal. No desk calculator could cope with this simple division problem. The reader may care to estimate how long it would take him, using pencil and paper, to determine *Phi* to 4600 places by performing a simple division in which there are 2300 digits in the divisor, 2300 digits in the dividend and 4600 digits in the quotient. The computer performed this gigantic calculation in *20 minutes*. It was then checked by inversion of the fraction. The digits of u_{n+1}/u_n are identical in every place with u_n/u_{n+1} except that the former begins with $1.6180\cdots$ and the latter with $0.6180\cdots$ The two ratios were found to coincide to 4598 decimal places. This is a vivid illustration of the accuracy and speed of working of a modern electronic computer.

A GEOMETRIC FALLACY

Another illustration of the connection between *Phi* and the Fibonacci series relates to an old geometric fallacy which is illustrated in figure 4.1. Construct a square whose side has a

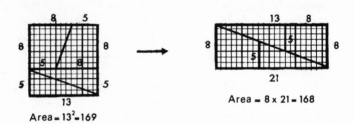

Fig. 4.1. Geometric fallacy (i)

length equal to the sum of two consecutive Fibonacci numbers. The figure shows $5 + 8 = 13$, but $21 + 34 = 55$ would do equally well. Dissect the square into the sections indicated and fit them together to form a rectangle. It is found that the areas of the square and the rectangle differ by 1 unit. Which is larger depends on the Fibonacci numbers selected. In the example of figure 4.1, the square is larger than the rectangle by 1 unit. If, however, instead of $5 + 8$ we had chosen $21 + 34$, we should have found that the rectangle was larger than the square by 1 unit. This corresponds to our discovery that consecutive ratios formed from an additive series are alternately greater than and less than *Phi*.

The explanation of this paradox is that the fit along the diagonals of the rectangles is not exact. Sometimes there is a gap of 1 square unit and sometimes an overlap of 1 square unit.

THE GOLDEN SERIES

But there *is* one additive series (and only one) which produces an exact fit. It is a series which makes use of the golden section. It may appropriately be called the *golden series*:

$$1, \phi, 1 + \phi, 1 + 2\phi, 2 + 3\phi, 3 + 5\phi, \cdots$$

If we construct a square the length of whose side is equal to the

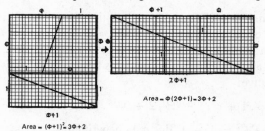

Fig. 4.2. Geometric fallacy (ii)

sum of any two consecutive numbers from this series, the fit will be exact: the areas of the square and rectangle will be equal. This is shown in figure 4.2. If we had chosen $1 + 2\phi$ and $2 + 3\phi$ as two consecutive numbers of the series, we should have found (remembering that $\phi^2 = \phi + 1$):

$$\text{area of square} = (3 + 5\phi)^2 = 55\phi + 34$$
$$\text{area of rectangle} = (5 + 8\phi)(2 + 3\phi) = 55\phi + 34$$

REMARKABLE PROPERTIES

The reader would have no difficulty in forming an additive series, that is, one which connects three consecutive terms according to the formula: $u_{n+1} = u_n + u_{n-1}$. Nor would he find it difficult to form a geometric series, i.e., one in which the ratio of any term to the preceding term was a constant for the series: u_{n+1}/u_n = constant for any value of n. He would form a geometric progression, such as 1, 3, 9, 27, 81, $\cdots$. But suppose the demand were to form a series possessing *both* these properties simultaneously? Not so easy!

There is, however, one such series, and only one. It is the golden series:

$$1, \phi, 1 + \phi, 1 + 2\phi, 2 + 3\phi, 3 + 5\phi, 5 + 8\phi, \cdots$$

This clearly possesses the additive property mentioned above. That it has also the second property follows from the fact that ϕ is a solution of the equation $x^2 - x - 1 = 0$, so that $1 + \phi = \phi^2$; It is easily deduced from this that the golden series may also be written thus:

$$1, \phi, \phi^2, \phi^3, \phi^4 \cdots$$

Each member of the series is positive. But since ϕ' is also a solution of the equation $x^2 - x - 1 = 0$, so that $1 + \phi' = \phi'^2$, there is a corresponding negative series having the same properties. It is an oscillating series, its terms being alternatively positive and negative:

$$1, \phi', 1 + \phi', 1 + 2\phi', 2 + 3\phi', 3 + 5\phi', 5 + 8\phi', \cdots$$

Because $\phi'^2 = \phi' + 1$, this may be written:

$$1, \phi', \phi'^2, \phi'^3, \phi'^4, \phi'^5, \phi'^6 \cdots$$

AESTHETIC CLAIM

The aesthetic appeal of the relations described in the few preceding paragraphs is necessarily reserved for those who have had some mathematical training. If the reader will glance back over these sections and recall the main points, he may be assured that his acquaintance with the muse of mathematics is but slight if he discovers no satisfaction at all either in observing that *Phi* is related not only to the Fibonacci series but also to any additive series whatever which is formed according to the same rule; or in the power of *Phi* to resolve convincingly the mystery of an ancient paradox; or in the simplicity and uniqueness of another of the properties of *Phi* expressed in the "golden series." But it may perhaps be taken for granted that the reader who has reached this point has resources for appreciation beyond those required for mere comprehension.

The conclusion might well be that the claim to beauty by the golden section and cognate topics has a man-made, a *purely artificial* basis, that its appreciation is an acquired taste, that the properties of this remarkable mathematical constant may have an interest for the mathematically wise and prudent but can never be revealed to babes. But the matter is not quite so simple as this. For we shall find (Chaps. XII and XIII) that the Fibonacci series, so far from being purely artificial, has connections with familiar natural phenomena and that the golden section, appropriately displayed, appears to have an immediate artistic appeal which demands no preliminary mathematical education. To amplify this, it is convenient to make a digression here and anticipate certain ideas discussed later and more fully in reference to the psychology of artistic appreciation.

THE DIVINE PROPORTION AND THE MAJOR SIXTH

Art appreciation is based on two distinct factors, one inherited and the other dependent on education, one from nature, the other from nurture. The former is instinctive, based in the racial unconscious as described by Jung (p. 177). The latter, the educative factor, is developed by training. Hunger is instinctive, but the taste for mother's milk, which is independent of any conscious education, can by training be developed into a partiality for sauerkraut or Gorgonzola cheese. As another example, consider the pleasure induced by a simple rhythmic beat. This, familiar even to babes and primitive men, involves no ingredient of conscious education, being part of the universal mental inheritance of the race, related possibly to the rhythmic motion experienced by the unborn child in the comfortable security of the womb. But the complex rhythmic pattern produced by expert African or Indian drummers demands training for its artistic understanding.

Now there are certain simple patterns both in mathematics and in music which demand only a minimum of artistic education for their appreciation as objects of beauty. In mathematics, the circle, the ellipse, the square; in music, simple musical intervals, can stimulate some emotional response with negligible preliminary training.

A rectangle, the lengths of the adjacent sides of which are in a ratio which is exactly or close to $\phi:1$, appears to afford a greater measure of satisfaction to a majority of people than do rectangles of different proportions. This was realized by the ancient Greeks, whose architecture incorporates features bearing witness to it. In recent times their observation has found empirical support in the experiments of the German psychologist Gustav Fechner (see Chap. V). Why this particular rectangle, which we may call "the golden rectangle," is preferred even to the square, or to the double square or to any other, is not understood. And when a matter is not understood; when, as in this case, no conceivable grounds for the preference are apparent; when a rational explanation is not even in sight; then scepticism concerning the facts arises. Accordingly, it is not surprising that many writers have dismissed the whole subject as nonsense. Nevertheless, it is difficult to believe that the alleged superiority of the golden rectangle, incorporated in ancient art, endorsed by Kepler, who wrote about "the divine proportion," and supported as it is by many modern experiments, is entirely without substance. It is wiser to regard this difference of opinion as just another example of the notorious difficulty of finding a rational explanation for aesthetic preferences. But the difficulty of accounting for a phenomenon does not invalidate its reality.

With some hesitation I venture to offer at this juncture a tentative explanation of the charm which the divine proportion holds for many people. It may appear to be far-fetched, but at least it brings a curious coincidence into relief.

What is the *immediate* experience of an observer confronted with the golden rectangle (Fig. 4.3)? However complex physiologically the act of seeing an object may be, the estimation by the eye of the relative lengths of the two adjacent sides of the rectangle is ultimately reducible to the instinctive measurement of the relative duration of two *time intervals*. What is subconsciously apprehended is the ratio of the time that would be required for the line of vision to swing from A to B to that of the time required to pass from A to D, these intervals being instinctively measured by one of the body's internal clocks. We may look at it in this way. While the focus of vision passes from *A* to *B*, a certain number of nerve impulses travel along the optic nerves in a period of time which is instinctively correlated with the muscular effort of the eyeball. In this way the experience of a *space interval* is reduced to the more immediate experience of a *time interval*. For example, the recognition that a figure is a square involves the realization, acquired by practice in early life, that the two time intervals involved are equal. Our problem is accordingly reduced to the

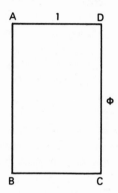

Fig. 4.3. Golden rectangle

question of why the ratio of the two time intervals corresponding to the lengths $AB:AD = \phi:1$ is a source of pleasure.

The observation offered here is that it may be related to the well-known fact that certain musical intervals are more acceptable by the mind than others because they are more harmonious. Three such intervals deserve special mention (excluding zero interval, i.e., two notes sounded in unison): the octave, the major third and the major sixth. These appear to be universally acceptable. Note that the last two intervals, having practically identical series of harmonics, are not always easily distinguished.

ESTIMATION OF PITCH

Now the estimation of pitch by the ear involves in some way the mental measurement of a time interval. However complicated physiologically the hearing process may be, when "middle c" (frequency $f = 256$ c/s) is heard, the essential fact is that a stream of pressure zones strikes the ear-drum and produces a corresponding stream of nerve impulses which are interpreted by the brain to correspond to time intervals of 1/256 second. Again, the *immediate* experience is of time intervals, measured against the body's internal clock, the length of the interval determining the pitch of the note. When the note which is an octave above middle c (c': $f = 512$ c/s) is sounded, the time interval is precisely one-half of that characteristic of middle c. Therefore it is not surprising that when an octave is sounded and there is complete agreement between the frequencies of the harmonics of both notes, the total effect is aurally acceptable.

Now the explanation given by Helmholtz of the harmonious blending of the tones of certain musical intervals was that an absence of "beats" between their harmonics resulted in *consonance*. The sound emitted by two notes such as those separated by a semitone is a *dissonance*: such an interval is rich in beats between interfering harmonics, a discord obnoxious to the ear.

Pythagoras noted the interesting fact that the musical intervals which are most consonant are reducible to the ratio of small integers:

INTERVAL	FREQUENCY RATIO
Unison	1:1 ⎫
Octave	2:1 ⎭
Major third	5:4 ⎫
Major sixth	8:5 ⎭

(The physics student will know that the matter is complicated by the presence of harmonics. The unison cannot be sounded without the octave being faintly audible, and the major third is rich in overtones scarcely distinguishable from those of the major sixth.)

We are now in a position to consider an hypothesis formulated to account for the artistic pleasure alleged to be derived from the golden rectangle. According to Helmholtz, a stream of aural nerve impulses from two notes sounded simultaneously produces consonance and aesthetic pleasure if the sound is free of beats produced by the fundamental tones or their harmonics. This occurs of course when two notes are sounded in unison. Now the visual centers of the brain are affected in an analogous manner when the eye beholds a square, which is a source of aesthetic satisfaction even if only by a stimulation, through association, of the aural response. A similar correspondence between aural and visual sensations, so that experience of the latter may, through associa- tion, evoke emotions called forth by the former, occurs when the ear hears an octave and the eye beholds a rectangle which is equivalent to a double square. But it is in accord with observation and experiment that the musical interval which gives the greatest satisfaction to the greatest number is the *major sixth*, frequency ratio 8:5. This corresponds to the pleasure experienced in seeing the golden rectangle, the adjacent sides of which are in the ratio ϕ:1, which is approximately equal to 8:5.

SUMMARY

We may summarize the foregoing argument as follows.

The assumption is made that certain nerve messages received by the visual centers of the brain can awaken associative echoes in the aural centers. There are three emotionally potent musical intervals which stand out from all others by virtue of their consonance: they are the unison, the octave and the major sixth. These are aestheti- cally pleasing because (according to Helmholtz) these pairs of tones produce no beats between their harmonics. Beats are characteristic of dissonance which offends the ear as discord. Corresponding to the three pleasing musical intervals are three congenial rectangles:

MUSICAL INTERVAL	EXAMPLE	RATIO OF FREQUENCIES	RECTANGLE	RATIO OF SIDE SEGMENTS
Unison	c:c	256:256 = 1:1	Square	1:1
Octave	c′:c	512:256 = 2:1	Double square	2:1
Major sixth	c′:e	512:320 = 8:5	Golden rectangle	8:5

Experience and experiment show that the most pleasing of these intervals are the (corresponding) major sixth and golden rectangle.

Now the immediate sense data received by the brain which is conscious of hearing a major sixth is the ratio of two time intervals, *viz.*, 8:5. Any cause, therefore, which produces visual consciousness of this ratio may result, either by association or in some other way, in inducing a pleasurable echo in the aural brain centers. The golden rectangle, with sides in the ratio 8:5, is just such a cause: hence its aesthetic appeal.

BINET'S FORMULA

The connection between the golden section and the Fibonacci series is seen from a new point of view by considering the general term of the series. This is Binet's formula:

$$u_n = \frac{1}{\sqrt{5}} \left(\frac{1 + \sqrt{5}}{2} \right)^n - \frac{1}{\sqrt{5}} \left(\frac{1 - \sqrt{5}}{2} \right)^n \qquad (n = 0, 1, 2, 3, \cdots)$$

No protracted calculation is required to show that this formula produces the first few integral members of the series

$$u_0 = 0, u_1 = 1, u_2 = 1, u_3 = 2, \cdots$$

With large values of n the second term of Binet's formula may be neglected. For instance, when n is no larger than 5, the formula gives (to the fourth place of decimals):

$$u_5 = 5.0403(2) - 0.0403(2)$$

The golden ratio has been shown to be

$$\phi = \lim_{n \to \infty} \frac{u_n + 1}{u_n}$$

Hence, ϕ is approximately equal to

$$\frac{1}{\sqrt{5}} \left(\frac{1 + \sqrt{5}}{2} \right)^{n+1} \bigg/ \frac{1}{\sqrt{5}} \left(\frac{1 + \sqrt{5}}{2} \right)^n$$

Thus $\phi = (1 + \sqrt{5})/2$, in the limit.

How close the approximation is, even for terms as early as u_{12}

and u_{13}, may be seen from the following values calculated from the formula:

$$u_{12} = 144.0014 \quad \text{(Fibonacci number, 144)}$$

$$u_{13} = 232.9991 \quad \text{(Fibonacci number, 233)}$$

THE LAW OF GROWTH

We are now in a position to understand that the claim to beauty at least in some areas of mathematics is not built on an artificial basis but grounded in the beauty of the natural world. For the law of biological growth, whether of a plant or an animal or any part of them, is an *exponential law*. A clear illustration of this is the growth of the shell of a mullusk. The radius r of curvature of the shell of a nautilus, for example, increases according to the mathematical formula

$$r = ae^{k\theta}$$

which is the polar equation of an equiangular spiral.

Now, Binet's formula for the discrete terms of the Fibonacci series can be regarded as a continuous function when n is large. It may be written

$$y = \frac{1}{\sqrt{5}} \left(\frac{1 + \sqrt{5}}{2} \right)^x = 0.4472 \times (1.6180)^x$$

For all practical purposes the Fibonacci numbers lie on this curve in its higher reaches.

Art and the Golden Rectangle

Still seeking the insight which can unveil the beauty latent in elementary mathematics, we shall in this chapter consider certain very simple geometrical figures, such as the golden rectangle, before we turn, in chapter VII, to more sophisticated examples which can evoke the aesthetic response. All the while we remain within the limits of topics cognate to the golden section, thereby exhibiting the fertility of even a very restricted field of mathematics. In all this we shall augment our anthology with further examples of interest and beauty. Most of these will confirm a claim already adumbrated, *viz.*, that education and training are the only means of developing and intensifying an aesthetic gift which is already inborn. For instance, the equiangular spiral (Fig. 7.6 on p. 101), which falls within our prescribed limits, requires a minimum of mental training before the fascination of its curve exerts its charm, but, when through education the appropriate mathematical key is applied, its hidden store of beauty yields further treasure to the mind.

"BEAUTY IS A WORD OF GOD"

Differences in education are mainly responsible for differences in taste; and the extent of intuitive (inborn) appreciation of beauty differs from person to person and from object to object. But the sentient creature who makes no response whatever to any lovely object; the individual, for example, who says "I hate flowers," is hardly human. Peter Bell is almost incredible:

> A primrose by the river's brim
> A yellow primrose was to him
> And it was nothing more.

Peter Bell's creator may very well tower over most of the human race in his aesthetic sensibility, as when he writes in ecstasy:

> To me the meanest flower that blows can give
> Thoughts that oft do lie too deep for tears.

But even he has scarcely begun to sound the depths of beauty which are to be uncovered in the humblest flower.

It was Tennyson who spoke the mind of the scientist, of the botanist who takes to the flower the dissecting tool and the microscope. Tennyson understood that its beauty knows no limits:

> Flower in the crannied wall
> I pluck you out of the crannies,
> I hold you here, root and all, in my hand,
> Little flower—but *if* I could understand
> What you are, root and all, and all in all,
> I should know what God and man is.

DEFINITIONS

The difference between the intuitive view of beauty—that of Wordsworth, and the analytical view of the scientist which Tennyson recognizes is expressed in the two briefest definitions of beauty which I have met. While no brief definition can be completely satisfactory, brevity is a merit in a definition. One is that of Thomas Aquinas: "Beauty is that which pleases in mere contemplation."[1] The other: "Beauty is a word of God."

The reader may perhaps feel that the quality and richness of a flower's beauty can scarcely be compared with the jejune harvest of the golden rectangle and the related topics of this chapter as amplified in chapter VII. Then consider the equiangular spiral, which has been incarnate in the nautilus sea shell almost since the birth of Time. Is not this one of the "words of God"? If it be so, then we must admit that, if the harvest it yields seems relatively scanty, it is because our tools are gross, our methods primitive and our understanding dull. Even the simple rectangular spiral of figure 5.5, progenitor of the equiangular spiral, is barren until a more than casual glance reveals properties like those set out in the list—a list of indefinite length—which is begun on p. 68.

To any reader of these pages who feels that mathematics is a barren wilderness, productive of only scanty material of aesthetic appeal, or who, through failure to search below the surface beauty, finds only superficial rewards, I venture to suggest that the fault may be his own. Francis Thompson, the poet, speaks in plain language to such:

> The angels keep their ancient places;—
> Turn but a stone and start a wing!
> 'Tis ye, 'tis your estrangéd faces,
> That miss the many-splendoured thing.

THE DIVINE PROPORTION

By restricting our field of exploration, confining it to the area of the "divine proportion," we shall endeavour to show that there is no need to roam widely to collect examples of mathematical beauty: they lie thick on the ground. These lines are written in the upper reaches of the Ötzal valley of the Austrian Tyrol, where, yesterday, in the space of a few hours, the trained eye of my wife found nearly fifty different specimens of wild flowers. With two or three exceptions, she tells me, these may be seen within a few hundred yards of our home in Somerset. It is the trained eye that sees.

In the remainder of this chapter we shall have the opportunity of examining simple figures lying near at hand, which appeal to the eye even of the non-mathematician, and then by a closer examination of these we shall find that they hide relationships, the discovery of which may become a source of pleasure to the mathematician of quite modest attainments. It is the laying bare of these unsuspected relationships with its small surprises and minor joys of achievement that constitutes part of the charm of mathematics. To take an example: you might not suspect, unless you had some acquaintance with the golden rectangle, that lopping off a square from it leaves a residue which is another golden rectangle, and that this process can be repeated indefinitely until a "point rectangle" is reached which has a unique situation. This may seem to be a trivial example, but it is simple and easily understood, and it makes the point that the unexpected results together with the pleasure of learning something hitherto unknown constitutes part of the charm of mathematics.

We shall give some space in the first instance to discussing the golden rectangle regarded as an aesthetically pleasing figure. Then its clear, simple appeal to the mathematical taste will be described.

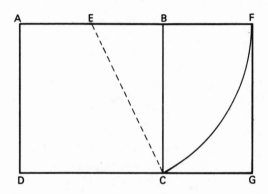

Fig. 5.1. Construction of golden rectangle

In chapter IV we anticipated the evidence that is now set forth in detail for the claim that a rectangle of certain proportions has an appeal to a wider population than a rectangle of any other shape.

The construction of the golden rectangle is a simple matter. The side AB of a square $ABCD$ is bisected in E (Fig. 5.1). With center E and radius EC draw an arc of a circle cutting AB produced in F. Draw FG perpendicular to AF meeting DC produced in G. Then $AFGD$ is the golden rectangle.

The proof is equally simple. Let $AB = 2$ units of length. Then $EC = EF = \sqrt{5}$ units. $AF/FG = (AE + EF)/FG = (1 + \sqrt{5})/2 = \phi$.

AF is divided by *B* in the golden section. *B* is sometimes called the "golden cut." It is associated with the idea of the "mean proportional": *AB* is the mean proportional of *AF* and *BF*:

$$\frac{AB}{BF} = \frac{AF}{AB}, \quad \text{i.e.,} \quad AB^2 = AF \cdot BF$$

EXPERIMENTAL AESTHETICS

Before we examine the geometrical properties of the golden rectangle, let us consider its claim to aesthetic merit.

As we saw in chapter II, Pythagoras believed that beauty was associated with the ratio of small integers. He experimented with the monochord and discovered that, when the length of the vibrating string was varied so as to emit the four notes of what today we call the "common chord," these lengths stood to each other in ratios expressible by small integers. For example, the octave ratio was $1:2$. The ratio that fascinated the Greeks was the golden ratio. This is related to the symbol of the Pythagorean brotherhood, the pentagram (Fig. 2.4 on p. 28), which contains several examples of the golden cut.

The proportions of the well-known Parthenon bear witness to the influence exerted by the golden rectangle on Greek architecture. The superstitious wise men of the Middle Ages, men of the breed of alchemists and astrologers, were fascinated by *Phi*. Kepler called it, as we have seen, the "divine proportion."

It is unfortunate that the golden section has attracted the enthusiastic attention of cranks. One of these measured the heights of 65 women and compared the results with heights of their respective navels, obtaining an average of 1.618. But when this enthusiast adds to this nonsense the claim that the accepted value of π (3.14159···) is incorrect and that it should be $6\phi^2/5 = 3.141608\cdots$, we recognise the type we are dealing with!

The most compendious work devoted to this topic, *Der goldene Schnitt* (1884) was by a German, Adolf Zeising. It appears to have stimulated the famous German psychologist, Gustav Fechner, to begin the first serious inquiry into the claims of the golden rectangle to have a special aesthetic interest. With characteristic German thoroughness, Fechner made literally thousands of ratio

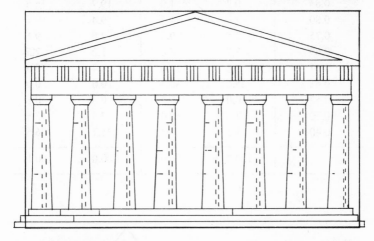

Fig. 5.2.   The Parthenon at Athens, built in the fifth century B.C., one of the world's most famous structures. While its triangular pediment was still intact, its dimensions could be fitted almost exactly into a Golden Rectangle, as shown above. It stands therefore as another example of the aesthetic value of this particular shape.

measurements of commonly seen rectangles—playing cards, windows, writing-paper pads, book covers—and found that the average was close to *Phi*. He also extensively tested personal preferences, and finally established that most people prefer a certain rectangle the proportions of which lie between those of a square and those of a double square.

Fechner's extensive experiments were made in 1876 and were rather crude. They were repeated by Witmar (1894), Lalo (1908) and Thorndike (1917). The results were similar in each case. The following tabulation of Fechner's and Lalo's measurements are of interest:

RATIO: WIDTH/LENGTH	BEST RECTANGLE		WORST RECTANGLE	
	Fechner, %	Lalo, %	Fechner, %	Lalo, %
1.00	3.0	11.7	27.8	22.5
0.83	0.2	1.0	19.7	16.6
0.80	2.0	1.3	9.4	9.1
0.75	2.5	9.5	2.5	9.1
0.69	7.7	5.6	1.2	2.5
0.67	20.6	11.0	0.4	0.6
0.62	**35.0**	**30.3**	**0.0**	**0.0**
0.57	20.0	6.3	0.8	0.6
0.50	7.5	8.0	2.5	12.5
0.40	1.5	15.3	35.7	26.6
	100.0	100.0	100.0	100.1

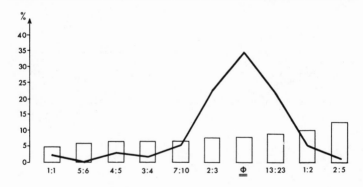

Fig. 5.3. Fechner's graph

If these figures mean what they seem to mean, they point un-ambiguously to a popular preference for a rectangular shape closely approximate to that of the golden rectangle. Of Fechner's observers, 75.6 per cent voted for it and of Lalo's, 47.6 per cent. The table is represented by the graph of figure 5.3.

THE GOLDEN ELLIPSE

Another shape of related interest is the "golden ellipse." This is the ellipse in which the ratio of the major to the minor axis is $\phi = 1.62 \cdots$. Here, as in the case of the golden rectangle, it has been shown empirically that the eye of the artist finds approximations to this ellipse more satisfying than others. For relishing pleasure in mathematical shapes, whether two- or three-dimensional, mathematical skill is unimportant. For the mathematician, however, such pleasure is a bonus.

Repeating his experiments with the ellipse, Fechner obtained the following results:

MINOR AXIS/MAJOR AXIS	PERCENTAGE
1.00	1.2
0.83	0.6
0.80	8.3
0.75	14.7
0.67	42.4
0.62	**16.7**
0.57	13.1
0.50	1.6
0.40	0.0

It will be seen that three observers out of four prefer an ellipse which is either the golden ellipse or so close an approximation as to be almost indistinguishable from it.

THE GERMAN "DIN"

It is interesting that the preferred shape of a rectangle approximates fairly closely to a shape which turns out to be economically useful. A German committee for standardization (Deutsche

Industrie Normen), in standardizing the shapes and sizes of paper sheets for printing, typewriting, and handwriting aimed at economy of paper. They minimized waste in cutting paper to smaller sizes through halving and halving again by choosing an original format that should remain similar in shape after bisection. A typist sheet, for example, when halved, should produce a geometrically similar sheet suitable for a personal letter. Only one rectangular form will achieve this. If the length and breadth of this standard size paper are x and y units, then after one folding,

$$x:y = \frac{y}{2}:x, \quad \text{i.e.,} \quad y = x\sqrt{2} \quad \text{or} \quad y \approx 1.4x$$

Thus, this rectangle is at once useful, economical and pleasing, for it is not seriously different in shape from the golden rectangle which is congenial to the artist's eye.

ADDITIVE SQUARES

The golden rectangle can be obtained approximately by another instructive method. Just as closer and closer approximations to the golden ratio are obtained from any additive series (of which the Fibonacci Series is an example) beginning with any two arbitrarily chosen terms (see p. 46), so closer and closer approximations to the golden rectangle result from an additive series of squares. In each case, the longer the series continues the closer is the approximation to the golden section and the golden rectangle respectively, regardless of the magnitude of the two initial terms. Let us begin the series with the two arbitrarily chosen squares, one shown shaded in figure 5.4. Then we add successively squares numbered 3, 4, 5, 6, 7, $\cdots$ and obtain the approximation to the golden rectangle shown. In figure 5.4 the ratio of the sides is $47/29 = 1.620\cdots$. But if we continue the process to square number 13, the approximation to *Phi* is closer: $521/322 = 1.6180\cdots$.

It will be clear from an inspection of the figure that the centers of successive squares lie on a spiral; and we shall find that this well-known spiral, the *logarithmic spiral*, is to be traced in other constructions involving the golden section. This is another example of a curve which, apart altogether from the neat simplicity

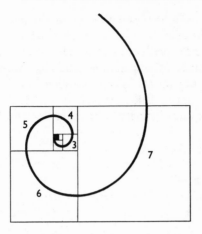

Fig. 5.4. Additive squares

of the mathematics involved, has an immediate appeal to the eye whether it be mathematically trained or not. Before we turn to this beautiful figure, however, let us examine the rectangular spiral (Fig. 5.5).

RECTANGULAR SPIRAL

Let OA (Fig. 5.5) represent unit length (ϕ'^0). At A erect a perpendicular to OA (AB) of length $1/\phi = 0.618034 \cdots$. For brevity

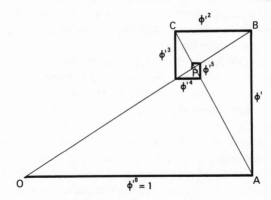

Fig. 5.5. Rectangular spiral

call this ϕ'. (Note: ϕ' is the symbol we have employed to denote the negative root of $x^2 - x - 1 = 0$, of which ϕ is the positive root.) At B draw BC perpendicular to AB and of length ϕ'^2, then CD perpendicular to BC of length ϕ'^3 and so on *ad infinitum*, producing a rectangular spiral, the coils of which diminish indefinitely to the limiting point P, called the *pole*.

This spiral has interesting properties, a few of which may be listed:

1. Turning points of even number all lie on OB; those of odd number all lie on BC.

2. OB and AC are mutually perpendicular.

3. The pole of the spiral, the limiting point P, is the intersection of OB and AC.

4. Each new arm of the spiral completes a triangle of which the other two sides are segments of OB and AC. These triangles are all similar, each being one-half of a golden rectangle.

5. The length of the spiral from A to P on the scale $OA = 1$, is *Phi*—a surprising conclusion!

To the interested reader other possible points of interest will suggest themselves. For examples, what is the position of the centroid? What is the radius of gyration of the spiral? What is the value of the ratio $OP:PB$?

These features of the rectangular spiral may serve to underline our point that aesthetic appreciation is consummated in two stages, the first through intuition, the second through education. Very few of the subjects tested by Fechner or Lalo could have been mathematically educated, so their choices must have been instinctive. The figure of the rectangular spiral would have an appeal to many, whether they were mathematically literate or not. But for its fuller appreciation, an understanding of such features as those listed above is required, and this comes only by education. To illustrate: it is natural to wonder what may be the total length of the spiral, measured from A inwards to the pole. It is obviously the sum of a series of segments which is convergent. But to discover that *it amounts precisely to Phi times OA* is to gain a glimpse of beauty in mathematics, and is a gratifying surprise to those whose taste for such revelations has been cultivated. That satisfaction is a specimen of the rewards of mathematical education.

In chapter VII when we meet the curvilinear spiral we shall find that these two stages of appreciation are even more distinctly differentiated.

CHAPTER VI

Beauty in Mathematics

There is ambiguity in this chapter heading. It may mean the pleasure derived from the mental activity which the study of mathematics generates; or it may mean the aesthetic feeling evoked by (e.g.) a mathematical theorem, which, on account of this feeling, is regarded as a thing of beauty. In what follows we are concerned with the latter meaning. An illustration may make the distinction clearer.

The study of a plane curve such as the parabola involves a succession of discoveries of hitherto unsuspected truths; each of these discoveries gives rise, in a greater or less degree, to an experience of beauty: the creation of harmony out of dissonance. The study is a pleasurable activity. Renan said, "There is a scientific taste just as there is a literary or artistic one," and we are thinking of beauty that has an objective ingredient.

AN EXAMPLE

Let us consider this curve from several different points of view.

i. In the first place, the parabola (Fig. 6.1) is a curve which is beautiful in itself. In the absence of any mathematical sophistication, merely to contemplate it is a pleasurable sensuous experience, though it might tax the wisdom of the psychologist to explain in what the pleasure lies. Of course, the curve is symmetrical about an axis, but so are many capital letters of the alphabet which can lay no claim to beauty. One might perhaps say that there is the tang of infinity about the curve as it journeys off into uncharted space, which is in contrast with the parochial quality of the region in the neighborhood of the focus *S*. But how much is this worth? The aesthetic appeal is not to be doubted, but its source is hidden. This is that part of the total artistic

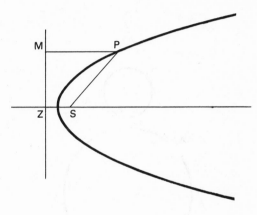

Fig. 6.1. Parabola

appreciation which is inborn; all the rest is acquired (see Chap. I, p. 10).

ii. Secondly, the parabola is a *locus* of great simplicity. It is the path traced out in a plane by a point moving in accordance with a simple law briefly stated thus: the point *P* is equidistant from a fixed point *S* (the focus) and a fixed line *ZM* (the directrix). If one tried to run to earth the origin of the aesthetic appeal of this, it would be found, in part, in the simplicity of the idea, the neatness of the method of generating a lovely curve. Moreover, the pleasure is enhanced with further education, for this allows a comparison between the parabola ($PS/PM = 1$) and the other conic sections, the ellipse ($PS/PM < 1$) and the hyperbola ($PS/PM > 1$).

iii. Since the marriage by Descartes (1596–1650) of geometry to algebra, it has been possible to represent the parabola in shorthand: $y^2 = 4ax$; this provides a powerful tool for revealing the properties of the curve. There is no appeal to the eye in this, but there is a great deal of aesthetic satisfaction arising from the application of coordinate geometry to the parabola. This may be confirmed by a reference to the section "*Phi* and the Parabola" in chapter VIII (p. 111).

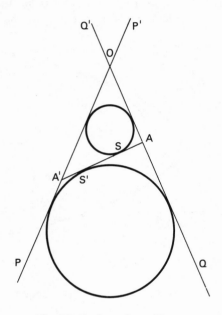

Fig. 6.2. Conic section: ellipse

iv. The viewpoint from which the parabola is seen in its most beautiful aspect is as a special section of a right circular cone. The most general conic section is an ellipse (AA', Fig. 6.2), which has two extreme forms—the circle and the parabola. In figure 6.2 $P'OP$, $Q'OQ$ represent generators of a right circular cone, AA' being a section of the cone by a plane which makes an angle with the axis of the cone greater than half the vertical angle AOA'. When this angle is equal to one half of AOA', the major axis AA' of the ellipse is parallel to the generator of the cone and is therefore of infinite length (Fig. 6.3). This extreme form of the ellipse is the parabola. The spheres inscribed to touch the cone in circles and the plane of the ellipse (Fig. 6.2) touch this plane at the foci *S*, *S'*. Moreover, the planes containing the circles of contact between the spheres and the cone intersect the plane of the ellipse in two lines which are the directrices. These exciting results are applicable, with modification, to the parabola (Fig. 6.3). To grasp

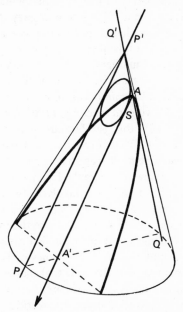

Fig. 6.3. Conic section: parabola (perspective)

these truths and their manifold implications is to glimpse beauty in mathematics.

v. The path of many a comet is a parabola with the sun at its focus. Each drop of spray from a water fountain describes a path which is a near-parabola. It is, in fact, a very elongated ellipse having the earth's center as one of its foci.

vi. Further interesting features of this lovely curve are noted in "*Phi* and the Parabola" in chapter VIII.

This example should serve to show the distinction between

beauty in mathematics and pleasure in searching it out. There is joy in rock climbing but it is not to be confused with the pleasure derived from viewing the scenery.

THE AESTHETICALLY UNGIFTED

I have spared little time hitherto to argue in these pages with those who doubt the reality of beauty in mathematics. They are, I believe, relatively few. But it may be necessary to say a word to those who think that aesthetic appreciation is rare, and these are many. I believe they are mistaken. Professor Hogben thinks that "the aesthetic appeal of mathematics may be very real for a chosen few." For these "few," he says, "mathematics exercises a coldly impersonal attraction." On an occasion when I was speaking to a Student Christian Movement meeting of sixth formers I happened to remark, incidentally, that the famous theorem of Pythagoras was "a thing of beauty." The explosion of derisive laughter that greeted this innocent remark was shattering. The reason for the outburst was, in my view, very simple. Every one knew that what I had said was true, but to admit it involved "wearing one's heart on one's sleeve" and this "isn't done" by sixth formers. One rarely hears the adjective "beautiful" from the lips of an adolescent: his private feelings are not for public display.

The universal popularity of board games with a mathematical basis is an argument against the view that mathematics is for the few. *Go* in Japan and chess in Russia are examples. Chess games and problems are found in many of the world's periodicals. It is relevant to our thesis to note that end-games are described as "beautiful," chess situations as "diverting," the check-mate as "neat," the solution of the problem as "elegant." Is there, then, beauty in chess but not in "mathematics"?

Supporting evidence for the widespread appeal of mathematics is found in the popularity of puzzles; in such fascinating columns as "Mathematical Games" published monthly for many years past in the *Scientific American* under the brilliant editorship of Martin Gardner; and in the dozens of books on "popular mathematics" which have been sold as paperbacks by the million.

In chapter I we discussed Beauty in its general context, as an introduction to a specialized form of beauty enjoyed by mathematicians, which is our present topic. Before examining the many illustrative examples of the next few chapters, in which we add to our small collection of examples of the golden section, it would be helpful to form a clearer idea of the nature of beauty in mathematics specifically. The task is not an easy one. As G. H. Hardy remarked, "It may be hard to define mathematical beauty, but that is true of beauty of any kind."

BEAUTY A GUIDE TO TRUTH

It is generally accepted that there is a close association in mathematics between beauty and truth. Indeed, there are those who would identify the one with the other, as does Keats in his "Ode on a Grecian Urn":

> Beauty is truth, truth beauty, that is all
> Ye know on earth, and all ye need to know.

And there is another point of view, that of Bertrand Russell:

In Mathematics we never know what we are talking about, nor whether what we are saying is true.

We are not now thinking of truth as a logical necessity—the *quod erat demonstrandum* conclusion which inexorably follows from stated premises. The truth of such a conclusion depends absolutely on the truth of the initial assumptions, provided that the logic of the argument is faultless. But there are truths that cannot be demonstrated, such as the truth of certain axioms and many propositions which, while there is reason to suppose them to be true, have so far surpassed the wit of man to establish. This is where mathematical beauty serves a useful purpose as a *guide* to truth.

We have seen that a supreme purpose of beauty vis-à-vis the human psyche is to serve as a stimulus to creative activity, which is one of the terminal mental satisfactions of man.

We inquired in chapter I whether aesthetic sensibility in man had any utilitarian purpose which would contribute to his survival in the struggle for existence in the course of his evolution. We could find no such purpose. That the feeling for beauty, however, can produce a mental ferment and generate new ideas in mathematics, and can serve as a guide to truth is an affirmation that many high ranking mathematicians endorse. Hadamard, for example, remarked that for the mathematician it was often the only criterion for deciding which method of attacking a problem would be successful:

[The] sense of beauty can inform us and I cannot see anything else allowing us to foresee. . . . This is undoubtedly the way the Greek geometers thought when they investigated the ellipse, because there is no other conceivable way.[1]

Perhaps the most extreme view of the pre-eminent importance of beauty in guiding one's thinking in science and mathematics was expressed by Dirac:

It is more important to have beauty in one's equations than to have them fit experiment.[2]

Hyperbole, no doubt, but perhaps the best way to enforce the lesson!

IDEAS IN POETRY AND MATHEMATICS

If we are to discover the source of the satisfaction that arises from the contemplation of a mathematical thesis, we shall do well to consider the more general question of the aesthetic pleasure associated with the creation and appreciation of great art as it is found in poetry or literature or music. Few (if any) have contributed more to the illumination of this question than the psychologist C. G. Jung in the development of his theory of the collective unconscious.

Let us consider again the example briefly mentioned in the Introduction: the first stanza of Gray's "Elegy Written in a Country Churchyard," reputed to be among the most popular poems in the English language:

> The curfew tolls the knell of parting day,
> The lowing herd winds slowly o'er the lea,
> The ploughman homeward plods his weary way,
> And leaves the world to darkness, and to me.

If a teacher of English literature were asked to account for the pleasure aroused by reading these lines, he would probably refer to the rhymes and rhythms, the tempo, the long, slow vowels, and the alliterative echoes. But would he say much about the content, the ideas, the imagery? Housman stated roundly that ideas in poetry are unimportant:

I cannot satisfy myself that there are any such things as poetical ideas.... Poetry is not the thing said, but a way of saying it.

If this is true, mathematical beauty must differ radically from that of poetry, for the working material of the mathematician is nothing but ideas. But is it true? It is certain that C. G. Jung would attach primary importance to the *ideas* conveyed by this poem. He quotes Gerhart Hauptmann, "Poetry means the distant echo of the primitive world behind our veil of words," and proceeds to amplify this by reference to his theory of the collective unconscious, which he distinguishes from the personal unconscious of the poet:

The collective unconscious is in no sense an obscure corner of the mind, but the all-controlling deposit of ancestral experience from untold millions of years, the echo of prehistoric world events to which each century adds an infinitesimally small amount of variation and differentiation.

If we recall Gray's primordial images of the ploughman, of the lowing herd and its winding way, of the progress of parting day leaving the world to darkness, we may see the relevance of Jung's ideas to his verse. Jung writes:

The man who speaks with primordial images speaks with a thousand tongues; he entrances and overpowers, while at the same time he raises the idea he is trying to express above the occasional and the transitory into the sphere of the ever-existing....

That is the secret of effective art. The creative process, in so far as we are able to follow it at all, consists in the unconscious animation of the archetype, and in a development and shaping of the image until the work is completed. The shaping of a primordial image is, as it were, a translation into the language of the present which makes it possible for every man to find again the deepest springs of life which would otherwise be closed to him.

Like the appreciation of music, pleasure in the pursuit of mathematics as a mental discipline springs from those deep layers of the human psyche which, having been developed in the early epochs of human evolution, lie buried beneath mental strata of later development. The accepted view that there is a definite connection between muscial appreciation and mathematical taste is based not only on the observation that many gifted mathematicians have had a warm appreciation of music (a few, like Einstein, having been skilled instrumental performers), but also on the similarity between the deep-seated structure of musical form and that of mathematical ideas. As Hardy said, "There are probably more people really interested in mathematics than in music." Perhaps many people enjoy music because they intuitively perceive its mathematical basis. As the conscious mind expresses itself in language and gesture, so the unconscious mind may become articulate in music and in mathematics. It is possible for neither music nor mathematics to assume any arbitrary form if it is to be intelligible to the mind. A fortuitous succession of notes makes as little sense as a haphazard chain of mathematical symbols. There is no more articulate language of the unconscious mind than music, but the syntax and the grammar of this language are not capricious; they are dictated in their broadest outlines by the texture and organization of the deep levels of the mind, which assumed its present structure in those aeons of evolutionary time that led up to the coming of *Homo sapiens*. So with mathematics. While we know something of prehistoric man's physical environment, and can speculate concerning the mental stresses which in the course of vast periods of time evolved a mentality of definite pattern functioning in a particular fashion, we know little or nothing of the reasons why this mentality should find satisfaction in certain types of mathematics rather than in a thousand other possible, unimagined types. The fact remains that the forms of music and the shapes of mathematics which appeal to our minds are directed by a basic mental structure which was itself an inexorable product of its terrestrial environment.

EVOLUTION OF AESTHETIC FEELING

Now we have reached the nucleus of the argument. The ultimate source of aesthetic sensibility to the various manifestations of beauty in mathematics is to be sought for in the unconscious mind or even (frequently) in the collective unconscious by virtue of which man is the heir of all the ages. The mental processes evoked in all men for a million years past by their physical environment have deposited a soil in which the roots of the psyche are deeply and securely implanted. Experiences of all the generations of a man's ancestors, repeated millions of times and recorded as memory structures in the brain, are scored ever more deeply as they are transmitted from generation to generation through the centuries, often showing a brief vitality in unaccountable dreams. The mind of a newly born baby is no *tabula rasa*: before he has had the opportunity to develop a conscious mind he is equipped with these inherited memory structures. For example, one of his earliest mental activities impels him to seek his mother's breast.

It is to the emotionally charged experiences of a thousand generations of our ancestors that we must look in order to discover the sources of aesthetic pleasure in art, in poetry, in music, in mathematics, and in other artistic forms. It is not impossible to guess what some of these experiences must be which, either because their repetition is so frequent or because they evoke strong mental excitement, have left their indelible traces on our mental structure; these traces are a fixed part of our human inheritance and the ground of our aesthetic appreciation.

MATHEMATICS AND MUSIC

The theory of the subconscious source of the appreciation of beauty in mathematics will be more readily understood if a comparison is made with the appreciation of music. This need not surprise us, for music is pre-eminently the language of the unconscious mind: in music the unconscious becomes articulate. Emotionally charged archaic memories, deeply buried, are easily aroused by melody and harmony, and it seems to be possible to relate certain primordial, racial experiences to familiar features of music.

Rhythm is basic to practically all music. No experience of the individual antedates that of the swinging motion of the womb as the mother walks. Is it not reasonable to associate appreciation of rhythm with the pleasure of the warm and comfortable prenatal days, and possibly also with the excitement, enjoyed by our simian ancestors for hundreds of generations, of swinging in the trees?

Or think of the many age-old experiences which produced emotional reactions in every member of the human race. Dancing, conversational exchanges, the sound of thundering hoofs of a stampede, flight from an enemy and many more such emotionally loaded experiences, are all readily stimulated by rapid, lively music which evokes the immemorial excitement.

Rallentando is a familiar feature of a piece of music which is approaching a final phase; the last chord, the tonic, is significantly termed the "home note." Is it unreasonable to relate the aptness of the *rallentando* to the universal emotional experience of hurrying home, then slowing down as sanctuary comes in view and finally halting on the threshold of home?

Again, long before man achieved the status of *Homo sapiens* he had on unnumbered occasions observed the flight of a stone in the earth's gravitational field—its rapid ascent, the long pause when it seemed neither to rise nor fall, followed by its accelerated descent. Surely the deeply buried memory of this oft-repeated experience is gently stirred when, a melody having reached its tonal climax, the high note is sustained for a moment or two before it descends to its level of departure?

These gentle stimulations of buried memories are common to many of the effects produced by music. And, of course, more often than not, several will operate simultaneously. The pleasure in rhythm does not exclude the excitement of *molto vivace* or the satisfaction of *rallentando*.

WHY IS RECOLLECTION PLEASURABLE?

But now the obvious question arises: why does the symbolic evocation of the memory of familiar racial experiences produced by music arouse a feeling of satisfaction which is an ingredient of aesthetic appreciation? The answer seems to be that it is precisely because these experiences *are* familiar. We are made to feel "at home." We enjoy a sense of security among things that we know and understand. The unfamiliar is unsettling. But any form of art, or poetry or music or mathematics, which runs smoothly and swiftly along well-worn memory grooves puts the mind at peace in a well-remembered environment. Ignore these age-old tracks and the aesthetic experience is impossible. Effective art depends upon primordial images.

In seeking to account for beauty in mathematics by reference to unconscious memories, as we have done for music, we must recognize an important difference: music is dynamic, mathematics is static. Nevertheless, the same source of aesthetic pleasure seems to operate in both. Mathematics will also evoke gentle, emotive memories, which, in the course of epochs, have scored deep into the unconscious mind. Because of the familiarity of these memories we understand and feel secure; we are "at home." This is admittedly speculative, but it does seem to be a normal condition of aesthetic satisfaction.

INGREDIENTS OF BEAUTY

Beauty in mathematics, as in music, is not elemental; it is a compound of several ingredients, which are not found in isolation—ingredients which have this in common: they stir buried memories which rise to awaken feelings in the conscious levels of the mind. Let us consider some of these. They are of a very general character, inherited by every member of the human race.

◘ ◘ ◘

The alternation of tension and relief is a universal emotion. In reading any form of serious mathematics, we experience alternately perplexity and illumination. Out of chaos comes order. Out of the many, the one. This, reaching the deepest levels of feeling, gently stimulates the aesthetic sensibilities. The effect is found in music when the alternation in (e.g.) a hymn tune of the dominant and the tonic—tension and relaxation—contributes to the beauty of the tune. Another example is the familiar one of resolution of discord into harmony.

In mathematics a student who finds himself bewildered by the wide variety of the series representations of such functions as e^x, $\log x$, $\cos x$, etc., is delighted to find a theorem like Taylor's which covers them all. "A beautiful generalization" may well be his reaction.

◘ ◘ ◘

The realization of expectation is a mental pleasure of more ancient standing than the human race itself. An example from music is the familiar sequence: dominant → tonic. An example from mathematics is found in the early paragraphs of this chapter, where the inscribed spheres depicted in figure 6.2, which touch the ellipse at its two foci, raise the question whether the same result would follow if the ellipse is projected into a parabola. The satisfaction that derives from seeing this expectation fulfilled (Fig. 6.3) is a spice which is relished in the flavor of the mathematical beauty of the conic sections.

◻ ◻ ◻

Surprise at the unexpected, conversely, is an emotion which we have in common with our animal ancestry. When a striking mathematical conclusion which has not been anticipated suddenly presents itself, old established emotions are stirred. An example might be the discovery of the Fibonacci series hidden in the Pascal triangle (Chap. X). Other examples are found in following pages.

◻ ◻ ◻

The perception of unsuspected relationships is another pleasurable experience old enough to have been built into our mental structure. One can imagine, for example, the excitement roused in the minds of primitive men when they first realized that there was a connection between the heights of the tides and the phases of the moon.

A simple illustration from music is the perception of the soprano melody in the tenor line of the Tallis Canon hymn tune:

An example from mathematics might be the relation between the equation of a conic:

$$x^2 + y^2 + 2gx + 2fy + c = 0$$

and its tangent at (x_1, y_1):

$$xx_1 + yy_1 + g(x + x_1) + f(y + y_1) + c = 0$$

Or there is the relation between the golden section and the 3–4–5 triangle (p. 43).

There does not appear, at first glance, to be any connection between the coefficients of the binomial expansion $(x + 1)^n$ as displayed in Pascal's triangle (p. 133) and the coefficients found in a formula for $\tan n\theta$. But consider the integers in the following:

$$(x + 1)^5 = x^5 + 5x^4 + 10x^3 + 10x^2 + 5x + 1$$

and compare with

$$\tan 5\theta = \frac{5 \tan \theta \quad\quad -10 \tan^3 \theta \quad\quad +\tan^5 \theta}{1 \quad\quad -10 \tan^2 \theta \quad\quad +5 \tan^4 \theta}$$

It turns out on further investigation that trigonometrical functions can be expressed algebraically without reference to right-angle triangles. This unification and generalization is a source of gratifying surprise. A feeling of increased mathematical power, too, comes by way of the remarkable formula

$$e^{i\theta} = \cos \theta + i \sin \theta \qquad (i = \sqrt{-1})$$

from which are derived

$$\cos \theta = \tfrac{1}{2}(e^{i\theta} + e^{-i\theta})$$

$$\sin \theta = \frac{1}{2i} (e^{i\theta} - e^{-i\theta})$$

This increase in our resources means, for example, that a problem in sines or cosines, such as $\int \sin^8 \theta \cdot d\theta$, can be changed into a more manageable one in exponentials.

◩ ◩ ◩

Mathematical beauty is found in *patterns*. The enjoyment of patterns is older than folk dancing. Hardy wrote:

> A mathematician, like a painter or a poet, is a maker of patterns. If his patterns are more permanent than theirs, it is because they are made with *ideas*.... The mathematician's patterns, like the painter's or the poet's, must be *beautiful*; the ideas, like the colours or the words, must fit together in a harmonious way....[3]

Dance tunes are patterns of rhythm and phrasing. In mathematics we can find a trivial example in the magic squares of chapter IX. A more serious example is the Chinese triangle (Chap. X). From matrices, groups, sets, we derive more modern illustrations offering scope for the exercise of the aesthetic faculty.

◻ ◻ ◻

"*Brevity is the soul of wit*." It may be the soul of beauty too. An example from poetry could be the brevity of the metre of Francis Thompson's "To a Snowflake":

> Fashioned so purely
> Fragilely, surely
> From what paradisal
> Imagineless metal
> Too costly for cost?

An example from mathematics might be Fermat's famous theorem, in which range and generality are condensed into a couple of lines.

Given x, y, z integers, the equation $x^n + y^n = z^n$ has no integral solutions if n is an integer greater than 2.

Goldbach's postulate may be quoted. It has been proved for all numbers less than 10,000:

Every even number is the sum of two primes.

An example from chapter III is:

If a series of integers is such that $u_{n-1} + u_n = u_{n+1}$, then $\lim_{n \to \infty} u_{n+1}/u_n = \phi$, the golden number.

An elementary example might be the proof of Pythagoras' theorem given by the Indian mathematician Bhāskara (born A.D. 1114). He simply draws four equal right-angle triangles as in figure 6.4. The area of each triangle is $ab/2$, so c^2 (the area of the

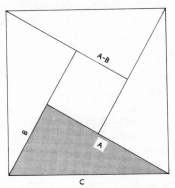

Fig. 6.4. Pythagoras' theorem

square) is equal to $4ab/2 + (a - b)^2 = a^2 + b^2$. The reader can easily verify the construction.

◘ ◘ ◘

"*Unity in variety*" was Coleridge's definition of beauty (Chap. I). It is frequently exemplified in music. As an example of the artistic quality of a mathematical theorem, consider the discovery by Johann Bernoulli (1667–1748) of the beautiful curve called the Brachistochrone (Fig. 6.5).

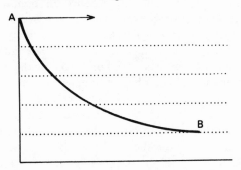

Fig. 6.5. Brachistochrone

A particle slides down a smooth curve from *A* to *B*. What curve makes the time of descent a minimum? Would it be a straight line, an arc of a circle (as Galileo supposed), or some other curve?

Bernoulli compared the path to that of a ray of light traversing a stratified layer of decreasing optical density (broken lines, Fig. 6.5); this is also a "least time" problem. He obtained the equation of the Brachistochrone:

$$y\left[1 - \left(\frac{dy}{dx}\right)^2\right] = \text{constant}$$

This is a cycloid, the curve described by a point on the circumference of a circle that rolls along a straight line.

The linking together of a problem in mechanics with a phenomenon in optics and relating the identical solution of both to a lovely curve—the cycloid, derived from pure geometry—has an artistic appeal that can scarcely be missed. As Polya remarks in this connection, "there is a real work of art before us."[4] It is unity in variety.

□ □ □

There is a *sensuous pleasure* to be derived from geometry. One of the gentle satisfactions enjoyed by all our ancestors, which must have left its mark on the unconscious mind, is the smooth sweep of the eye along the many quiet curves found in Nature. The smoothness of their contours is associated with the ease and comfort of the eye's muscular effort. Jagged and jerky lines have been shown by psychologists to produce an opposite mental effect. The curves that the human gaze has followed for a million years include the sea horizon, the skyline of the rolling downs, the rainbow, the meteor track, the parabola of the waterfall, the slingstone and the arrow, the arcs traced in the sky by the sun and the crescent moon, the flight of a bird, and many others.

Such purely sensuous pleasure is an ingredient of the aesthetic joy found in the geometry of the circle, the ellipse and other conic sections, as well as of the cycloid, the catenary, the graphs of trigonometrical functions, the cardioid, the logarithmic spiral ($\log \rho = a\theta$), Archimedes' spiral ($\rho = a\theta$), the limaçon and many other lovely shapes.

A melody may mirror such grace. It is rare to find the jagged contour of widely spaced notes in a melody. The melodious phrase may ascend and descend gently, register minor and major climaxes, pirouette like a ballerina, before subsiding smoothly to its point of departure. Such primeval aesthetic satisfactions, mathematical and musical, are rooted in the racial unconscious of humanity.

◨ ◧ ◨

A sense of wonder, even of awe, in the presence of the infinite, is one of the basic human emotions. Through all the aeons of time when man has stood beneath the cold light of stars and gazed into the unbounded depths of space; and especially since man first understood, a century ago, that an age-long stretch of evolutionary history lies behind him, infinity has been for him an emotionally charged concept. Music has power to arouse this emotion. So has mathematics. A divergent series of any sort induces this sense of infinity even as a convergent series leads to the related idea of the infinitesimal. Both feelings are roused by the spectacle of the curve of the hyperbola streaking off to infinite distance, simultaneously reducing its separation from its asymptote without ever reaching it. These are aspects of the aesthetic experiences of mathematics which easily pass unnoticed as such.

With this we may associate the baffled sense of mystery produced by certain mathematical theorems, the beauty of which is accompanied by an initial feeling of inadequacy to explain such remarkable results.

An example is Pascal's "Mystic Hexagram":

If a hexagon is inscribed in a conic, then the intersections of the three pairs of opposite sides are collinear (Fig. 6.6).

A beautiful theorem! Pascal (1623–1662) proved it when he was only sixteen years old and gave the figure its name.

Brianchon proved a theorem as follows:

If a hexagon is circumscribed about a conic, then the joins of the three pairs of opposite vertices are concurrent.

Many theorems of this type are found in treatises on projective geometry. The trained minds of mathematicians have delighted in their beauty for centuries past.

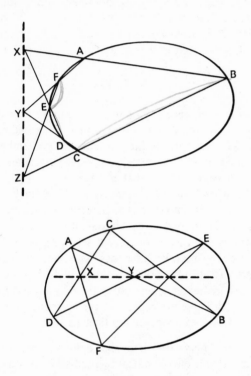

Fig. 6.6 "Mystic hexagram"

THE SEEING EYE

"The trained minds": the enjoyment of beauty in mathematics is for the most part an acquired taste. The eye has to be educated to *see*. How much of beauty the eye misses for lack of training! "Having eyes, they see not." Even the most highly trained mathematician must remain unmoved by much of the splendor because it is hidden from his keenest sight. "I can't see much in your scenery here," said an American tourist to a guide in Wordsworth's country. "Don't you wish you could, sir?" was the apt retort. Did anyone ever "see" more than Wordsworth? We may well doubt it. What we may never doubt is that there is more to be seen. The point has been well made by Sir Francis Younghusband. Moved by the beauty of Kashmir scenery he wrote:

There came to me this thought, which doubtless has occurred to many another beside myself—why the scene should so influence me and yet makes no impression on the men about me. Here were men with far keener eyesight than my own, and around me were animals with eyesight keener still.... Clearly it is not the eye but the soul that sees. But then comes the still further reflection: what may there not be staring *me* straight in the face which I am as blind to as the Kashmir stags are to the beauties amidst which they spend their entire lives? The whole panorama may be vibrating with beauties man has not yet the soul to see. Some already living, no doubt, see beauties we ordinary men cannot appreciate. It is only a century ago that mountains were looked upon as hideous. And in the long centuries to come may we not develop a soul for beauties unthought of now?

CHAPTER VII

Simple Examples of Aesthetic Interest

In this and the following chapter we consider specimens of simple mathematical exercises which illustrate how elegantly and how unexpectedly the golden section and its related topics emerge from a wide variety of problems—practical, algebraical, and geometrical.

Readers of these pages may wonder why I did not dish up a mixed grill of famous mathematical theorems of acknowledged aesthetic attraction. There are many such. The explanation is simply that, by drawing from a terrain of mathematics which is at once elementary and circumscribed—confined, in fact, to one main topic—I hope to convince those whose mathematical skill is limited that, even in so restricted an area as we are concerned with, examples of mathematical beauty.are not difficult to find. In the next chapter we shall consider items of the anthology which make a greater demand upon mathematical knowledge than is required in the present chapter, but they have the same objective—to nourish and develop a taste for beauty in mathematics.

We begin with a practical problem. A chemist in an effort to reduce the labour of his research finds that a knowledge of the Fibonacci series enables him to assess it *a priori*.

Then we pose a problem reminiscent of the famous right-angle triangle theorem of Pythagoras; the answer resurrects the golden section.

This is followed by another triangle problem with a similar result.

After looking briefly at the Cross of Lorraine, we come to a three-dimensional parallel to the golden rectangle—the golden cuboid—which proves to be rich in instances of the "divine proportion."

Finally, we touch on the logarithmic spiral, a truly beautiful addition to our theme, which is more fully treated in chapter XIII.

THE PILL MAKER

A manufacturing chemist, looking for the best combination of likely ingredients for a pill for therapeutic purposes, experimented in the following systematic, labor-saving manner.

The first recipe included all of 9 ingredients. The second included 8 ingredients taken 7 at a time—8 recipes. The third included 7 ingredients taken 5 at a time—21 recipes. And so on, in accordance with the following table:

Number of ingredients available:	9	8	7	6	5
Number included in recipe:	9	7	5	3	1
Number of possible recipes:	1	8	21	20	5

The total of 55 recipes is a Fibonacci number (u_9).

If the chemist had used 13 ingredients on the same labor-saving plan, the tabulation would have been:

Number of ingredients available:	13	12	11	10	9	8	7
Number included in recipe:	13	11	9	7	5	3	1
Number of possible recipes:	1	12	55	120	126	56	7

The total is 377, again a Fibonacci number (u_{13}).

The more advanced mathematician will observe that the foregoing are particular examples of a general rule which can be deduced from the Pascal triangle (p. 134).

A TRIANGLE LIMITED BY THE GOLDEN RATIO

The cosine formula of elementary trigonometry expresses the relation between the area of the square erected on one side of a triangle, the sum of the areas of the squares erected on the other two sides, and the angle θ contained by these two sides:

$$QR^2 = PQ^2 + PR^2 - 2PQ \cdot PR \cos \theta$$

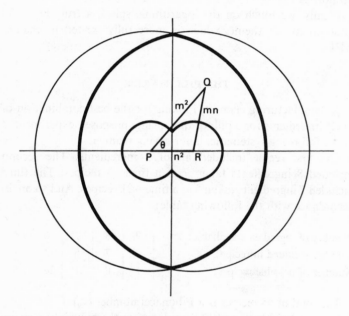

Fig. 7.1. Triangle limited by the golden ratio

and, of course, if θ is a right angle we obtain the special case of Pythagoras' theorem (Fig. 7.1).

Suppose now we examine the relation between the sides of the triangle under different conditions, *viz.*, that the area of the square one on side of the triangle is equal to the area of the rectangle contained by the other two sides, i.e.,

$$QR^2 = PQ \cdot PR$$

Letting $PQ = m^2$, $PR = n^2$, and $QR = mn$, then, since $PQ < PR + QR$, $m^2 < n^2 + mn$; that is:

$$\left(\frac{m}{n}\right)^2 - \frac{m}{n} - 1 < 0$$

Factoring,

$$(m/n - \phi)(m/n - \phi') < 0 \quad (\phi = 1.6180\cdots, \quad \phi' = -0.6180\cdots)$$

The first factor is negative, the second positive.

It follows that, if PR is of unit length ($n = 1$), then, since $\phi' < m/n < \phi$ or $\phi'^2 PR < PQ < \phi^2 PR$, the point Q must lie within the area bounded by the outer and inner circular arcs of figure 7.1 denoted by the heavy lines. In the figure the concentric circles with center P have radii ϕ^2 and ϕ'^2; similarly with concentric circles with center R.

A TRIANGLE INSCRIBED IN A RECTANGLE

The following problem published in a mathematical journal, is an example of the fact that many of the elegant results obtained can, if one penetrates a little more deeply into the solution, be made to yield additional satisfaction. The solution provides the element of surprise we have noted before, and, with its supplementation, it is neat and satisfying.

PROBLEM—*To inscribe a triangle within a given rectangle so that the three triangles thus formed are of equal area.*

Let $ABCD$ be any rectangle (Fig. 7.2). Take a point P in CD and a point Q in BC. We inquire: what are the positions of P and Q if $\triangle ABQ = \triangle PCQ = \triangle ADP$?

Let $DP = a$, $PC = b$, $BQ = c$, $QC = d$.

Since the triangles are equal in area, $c(a + b) = bd = a(c + d)$, whence $bc = da$. Eliminating a, $bd/(c + d) = bc/d$, i.e., $d^2 - cd - c^2 = 0$ or $(d/c)^2 - d/c - 1 = 0$, so that $d/c = \phi$. Similarly, $b/a = \phi$.

Thus, the problem is solved by dividing BC, CD in the golden section.

The solution, as published, stopped at this point. But it soon becomes clear that interesting corollaries can be deduced.[1] In the

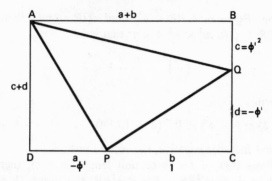

Fig. 7.2. Triangle inscribed in a rectangle

present instance, a natural question might be: what is the condition that $\triangle PAQ$ is isosceles?—say, that $PA = PQ$?

Evidently it is that

$$b^2 + d^2 = (c^2 + d)^2 + a^2$$

Since $d = c\phi$ and $b = a\phi$, this gives

$$a^2\phi^2 + c^2\phi^2 = c^2(1 + \phi)^2 + a^2$$

i.e.,

$$\frac{a^2}{c^2} = \frac{1}{\phi} + 2 = \phi^2 \quad \text{and} \quad a:c = \phi:1$$

Thus, if $b:a = a:c = \phi$, triangle PAQ is isosceles, and $a = d$. Since $d = c\phi$, and $b = a\phi$,

$$\frac{a + b}{c + d} = \frac{a(1 + \phi)}{c(1 + \phi)} = \phi$$

Hence, the condition that $\triangle PAQ$ is isosceles is that the rectangle is a golden rectangle. This is a worth-while result—shall we say, a pretty result? The initial solution is enriched, and we are led on to inquire whether any further enrichment is possible.

Further exploration shows that, $ABCD$ being a golden rectangle, $\angle APQ$ is a right angle.

If we take the definite case in which $AB = \phi$ and $BC = 1$, then

$$a = -\phi', \quad b = 1, \quad c = \phi'^2, \quad d = -\phi'$$

The proofs are left as an exercise for the reader.

THE CROSS OF LORRAINE

We have referred to the ubiquity of the golden ratio, to the fact that it crops up unexpectedly in strange places, producing the kind of feeling of pleased surprise which we experience in meeting an old friend. As another illustration of this, consider a problem publicized by Martin Gardner, the able editor of books of mathematical games and puzzles. It is related to the Cross of Lorraine, made famous throughout the Western world by General Charles de Gaulle.

The cross with its two transverse beams is represented in figure 7.3 where it covers on squared paper an area of 13 major squares each of unit area. The problem is to draw a straight line

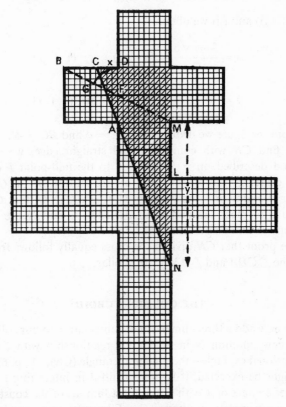

Fig. 7.3. Cross of Lorraine

through the point A in such a way as to divide the area of the cross into two equal parts. It may be solved either by calculation or with compasses and straight edge.

Let the required line be CN, cutting the outline of the cross in C and N (Fig. 7.3). Let $CD = x$, $MN = y$.

Since the cross covers 13 unit squares, the area to the right of CN is required to total $6\frac{1}{2}$ squares, so that the shaded area must amount to $2\frac{1}{2}$ squares.

Since $\triangle CDA$ is similar to $\triangle AMN$,

$$\frac{x}{1} = \frac{1}{y}, \quad \text{i.e.,} \quad xy = 1 \qquad \text{(i)}$$

$$\text{Shaded area} = (x + 1)(y + 1) = 2 \times 2\tfrac{1}{2} \qquad \text{(ii)}$$

From (i) and (ii) we obtain

$$x = \frac{3 \pm \sqrt{5}}{2} = (1 + \phi) \quad \text{or} \quad (1 + \phi')$$

$$y = \frac{3 \pm \sqrt{5}}{2} = (1 + \phi') \quad \text{or} \quad (1 + \phi)$$

From the figure we deduce that $NL = \phi$ and $BC = \phi'$.

To find CN with compasses and straight edge, we use the method described on p. 27. Join B to the mid-point F of AD. With center F, radius FD, draw an arc cutting BF in G. With center B, radius BG, draw an arc cutting BD in C. Join CA and produce to meet the outline in N. Then C and N are golden sections of unit lengths, and $NL = \phi$, $BC = -\phi'$.

The proof that CN divides the cross equally follows from the fact the $\triangle CDA$ and $\triangle AMN$ are similar.

THE GOLDEN CUBOID[2]

We now add a three-dimensional contribution to our collection. This new addition is in analogous relationship with the two-dimensional example—the golden rectangle (Chap. V, p. 61), but, as might be expected, it is more fruitful in interesting features. A brief contact of ϕ with π, the most famous of the constants of pure mathematics, is made; the Fibonacci sequence is again encountered; and other results of attractive simplicity are obtained.

This example affords an opportunity to remind the reader of a matter discussed briefly in chapter I. No one who has had experience of it can doubt the reality of the joy of creative activity, or of making an original discovery, however trifling. This does not necessarily involve revealing new knowledge but of *finding for oneself* something fresh, even though it had been known for a thousand years. Very few, if any, young students of mathematics, are entirely without this happy experience, and some have regarded this as one of the main ends of the education of man—"his vocation is to be a creator" (see p. 21).

The present writer in this simple instance renewed once more the pleasure that has often been his when, while he concentrated on a topic of mathematical interest, some "new thing" has swum into the field of his mental vision. That the series of golden cuts applied repeatedly to the golden rectangle, resulting in an unending sequence of similar rectangles (Chap. V, p. 61), could be paralleled in the analogous case in three dimensions was of course a very minor "discovery"; nevertheless, it provided a taste of the unique joy of creative activity of a sort which can become the experience of anyone who will take a little trouble to dig beneath the surface of mathematics.

But the main point I wish to emphasize here is not the satisfaction imparted to the discoverer by his discovery, but rather the multiplication of his solitary pleasure as he, by showing his "new thing" to others, enlarges the vision of those who are able empathetically to share his creative joy. Is this not, effectively, what the artist, the mathematician and the scientist achieve on behalf of mankind—a widening and clarifying of mental vision? A new insight into truth, a new specimen of beauty, once it is recorded and shared, is not lost but added to the growing store. Civilization, as the centuries pass, is becoming ever more richly endowed with beauty of an ever-widening range—from a cave painting to a Rembrandt, from a Greek temple to Coventry cathedral, from African drumming to a Beethoven symphony.

The ancient Greeks uncovered many lovely things in the realm of mathematics, but the store of beauty in this discipline has today become immeasurably greater. It is true that the greater part of this demands considerable mathematical training for its appreciation, but there remains a large and growing area—an area, moreover, without limits—open to exploration by the mathematician of humble attainments. The simple specimen which follows illustrates this point.

Consider the problem of finding the dimensions of a cuboid (rectangular parallelepiped) of unit volume which has a diagonal 2 units in length.

Suppose the lengths of the edges are a, b, and c. Then:

$$a \cdot b \cdot c = 1 \tag{i}$$

$$\sqrt{a^2 + b^2 + c^2} = 2 \tag{ii}$$

If only the ratios of these lengths are required, we may without

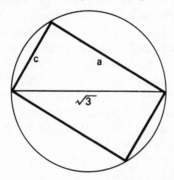

Fig. 7.4. The golden cuboid: base

loss of generality write $b = 1$, provided that $a \cdot c$ can have the value unity and that $a^2 + c^2 = 3$. Now it is evident from figure 7.4, which represents the base of the cuboid of unit height, that $a \cdot c$ has a maximum value when $a = c = \sqrt{3/2}$, so that $a \cdot c$ may have any value from zero to 3/2.

From (i) we have $c = 1/a$; substituting in (ii) we obtain

$$a^2 + 1/a^2 = 3, \quad \text{i.e.,} \quad a^4 - 3a^2 + 1 = 0,$$

whence

$$a^2 = \frac{3 + \sqrt{5}}{2} = 1 + \phi = \phi^2$$

so that $a = \phi$, the golden section. This, as we have seen, is the positive solution of the equation $x^2 - x - 1 = 0$ and has the value u_n/u_{n-1} as $n \to \infty$, u_n being a number of the Fibonacci series.

Since from (i) $c = \phi^{-1}$, we find the required ratios to be

$$a:b:c = \phi:1:\phi^{-1}$$

It is easily verified that $\sqrt{\phi^2 + 1 + \phi^{-2}} = 2$, the diagonal. Moreover, it turns out that these are the dimensions of a cuboid of unit volume: $\phi \times 1 \times \phi^{-1} = 1$.

Here, then, is another example of *Phi* appearing out of the blue! No one meeting this simple problem would have guessed

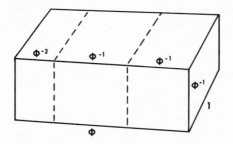

Fig. 7.5. The golden cuboid: perspective

that the solution would involve the golden section. We meet again one of the ingredients of the beauty of mathematics described in chapter V, surprise at unexpectedly meeting an old friend!

As we inquire further into the properties of this solid, we shall discover other features which justify our naming it the *golden cuboid*.

1. It is clear from figure 7.5 that both the lengths of the edges and the areas of the faces are in geometrical progression:

$$\phi^{-1}:1:\phi = 1:\phi:\phi^2$$

2. Four of the six faces are golden rectangles.

3. While its volume is that of a unit cube, the total surface area of the golden cuboid is $2(\phi + 1 + \phi^{-1}) = 4$.

4. The ratio of the area of a sphere circumscribing it to that of the cuboid is π—an interesting result.

We have mentioned here four of those features, but diligent digging might well extend the list. For example, we might wonder what the radius of gyration of this cuboid about an axis of symmetry might be; or, perhaps, what would be the volume of the symmetrically inscribed ellipsoid; and other such inquiries, any of which might conceivably produce a pleasant surprise. The limit to this seems to be set by our lack of imaginative power.

One further point is of interest.

We have seen (Fig. 5.1, p. 61) that if a square is cut off from a golden rectangle, the figure that remains is a golden rectangle ϕ^{-2} times the area of the original and that, of course, this dissection may be repeated an indefinite number of times with the same result. Now, if *two* cuboids of square cross-section $(\phi^{-1} \times \phi^{-1})$ are cut off from the golden cuboid (broken lines, Fig. 7.5), the lengths of the edges of the remaining cuboid are in the same ratio as those of the original cuboid, so that this is also a golden cuboid, ϕ^{-3} times the volume of the original.

The repetition of the decapitation process will lead to an indefinitely small cuboid enclosing a limiting point. The determination of the location of this point is offered as an exercise to the reader.

THE LOGARITHMIC SPIRAL

We come finally to one of the most beautiful of mathematical curves. It is known as the *logarithmic spiral*. For a reason which will appear later (Chap. XIII) it is also called the *equiangular spiral*. We shall learn that these spirals have been of common occurrence in the natural world for millions of years.

We shall find that, in studying this *spira mirabilis*, we are not transgressing the declared limits of this anthology, since the golden section and the pentagram (Chap. II) of Pythagoras and the Fibonacci series (Chap. IV) of Leonardo of Pisa are all associated with this remarkable curve.

This elegant spiral will make an appeal first to our artistic sensibilities and only second to our sophisticated mathematical appreciation. One wonders whether this prior appeal of the curve's form is related to the fact that it has been a familiar sight in the world of Nature since there were men to see it.

We begin with an intriguing property of the golden rectangle, introduced in chapter V (Fig. 5.1). Given such a rectangle $ABCD$ (Fig. 7.6), in which $AB:BC = \phi:1$; through E, the golden cut of AB, draw EF perpendicular to AB cutting off from the rectangle the square $AEFD$. Then the remaining rectangle $EBCF$ is a golden rectangle. If from this the square $EBGH$ is lopped off, the

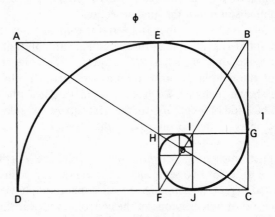

Fig. 7.6. Logarithmic spiral

remaining figure $HGCF$ is also a golden rectangle. We may suppose this process to be repeated indefinitely until the limiting rectangle O, indistinguishable from a point, is reached.

The following features of interest in this figure should be noted:

1. The limiting point O is called the *pole* of the equiangular spiral which passes through the golden cuts D, E, G, J, $\cdots$. We shall see later (p. 172) that the general equation of this spiral is

$$r = ae^{\theta \cot \alpha}$$

(The sides of the rectangle are nearly but not quite tangential to the curve.)

2. Alternate golden cuts on the *rectangular* spiral $ABCFH \cdots$ lie on the diagonals AC and BF. This suggests a convenient method of constructing the figure.

3. The diagonals *AC* and *BF* are mutually perpendicular.

4. The points *E, O, J* are collinear, as also are the points *G, O, D*.

5. The four right angles at *O* are bisected by *EJ* and *DG* so that these lines are mutually perpendicular.

6. $AO/OB = OB/OC = OC/OF = \cdots$. There is an infinite number of similar triangles, each being one-half of a golden rectangle.

The first of the six features shows the connection between the logarithmic spiral and the golden section.

The relation to the Fibonacci series is evident from the figure, for the spiral is seen to pass through diagonally opposite corners of successive squares such as *DE, EG, GJ,* $\cdots$. The lengths of the sides of these squares form a Fibonacci series. If the smallest square shown in figure 7.6 has a side of length *d*, the adjacent square has side of length *d* also, the next square has side of length 2*d*, the next 3*d* and so on, giving the series 1*d*, 1*d*, 2*d*, 3*d*, 5*d*, 8*d*, $\cdots$.

Another interesting property of the spiral is worth noting. However different two segments of the curve may be in *size* they are not different in shape. Suppose a photograph were taken with the aid of a microscope of the convolutions near the pole *O*, too small to be visible to the unaided eye. If such a copy were suitably enlarged it could be made to fit exactly on a spiral of the size of figure 7.6. The spiral is without a terminal point: it may grow outwards (or inwards) indefinitely, but its shape remains unchanged.

This is one of the points of contact between the Fibonacci series and the world of nature. The successive chambers of the nautilus sea shell (*frontispiece*) are built on a framework of a logarithmic spiral. As the shell grows the size of the chambers increases but their shape remains unaltered.

CHAPTER VIII

Further Examples

In the previous chapter we presented simple examples that required only elementary mathematical attainments for their appreciation. In this chapter we have the same end in view, but rather more advanced mathematics will be involved. We adhere to our plan to draw illustrations from the limited field of the golden section. This makes the point that needs emphasis, that one has not to roam very far to find interesting specimens: they often present themselves for inspection. Since four-fifths of the results obtained in this chapter are original creations of the last twelve months, it is probable that they are new to elementary mathematics; this helps to justify the claim that the limits are not over-restrictive.

All but one of the examples in this chapter are from plane and solid geometry. It is often said that the teaching of geometry is "on the way out," to make room for "modern mathematics"—sets, groups, matrices, etc. This is probably unavoidable, but none the less regrettable. Some of the loveliest gems of elementary mathematics lie in the realm of geometry. However great may be the utilitarian value of the binary notation and computers, they are no substitute for the aesthetic worth of such disciplines as advanced plane geometry and conic sections.

This raises a question of some practical importance. Should mathematics be studied with a utilitarian objective in mind or with a view to its aesthetic values? Both are important, but for school children and university students, I would give the latter priority. There is today a strong and growing movement to make mathematics *interesting* to school children, in the hope of reducing numbers of those who say "I *hate* mathematics!" and of increasing the number of mathematicians essential for meeting the needs of industry. These are commendable objectives, but I would aim higher than this. Mathematics should appeal to adolescents not only because it interests them but also because it satisfies their sense of beauty. "Whatsoever things are lovely ...," wrote St. Paul, "think on these things."

In an essay published some years ago in *Universitas*, a journal of the University of Ghana, I wrote under the title "The Motive of the Scientist":

It seems that it is the aesthetic motive, partly or wholly subconscious, which is the inspiration of the man who is the truly representative scientist. No evanescent or superficial interest accounts for the patient quest and dazzling trophies of natural philosophy. Some puissant motion in the deep places of the spirit is required to match the passion and the splendour. Joy in created form, the eye for unworldly perfection, the response which insight makes to self-authenticating truth—such are more commensurate with the beauty that is made manifest.[1]

To elevate the argument to its highest plane, one might quote the Westminster Catechism:

QUESTION—What is the chief end of man?
ANSWER—To glorify God and enjoy Him for ever.

Is this too "elevated" for children? If, as has been said, Beauty is a word of God, should not youth be taught this language? Because he has spoken his native tongue for a decade or more, the child in quest of aesthetic pleasure will go straight to the school classroom where poetry and literature or music are taught. But, if only he had learned the Creator's vocabulary, he would make a beeline for the classroom where science and mathematics are taught, where the feast of beauty is unlimited both in abundance and variety. Every discerning teacher knows that in the sphere of the created world, Beauty is an utterance of the divine voice, but scarcely one in ten thousand attempts to teach this language to the rising generation: "It's not in the syllabus."

□ □ □

TRISECTORS OF AN ANGLE

We have seen that there is a connection between *Phi* and the Fibonacci series, and that there is also a connection between *Phi* and the trigonometrical functions of $\pi/20$ radians. Both of these features are found in the next elementary problem.

Problem: The trisectors of an angle 3θ divide a straight line into three segments of lengths equal to three members of the Fibonacci series, u_n, u_{n+1}, and u_{n+3}. Find the limiting value of θ as n approaches infinity.

As n tends to infinity $u_{n+1}/u_n \to \phi$ and $u_{n+3}/u_n \to \phi^3$. Set out in a straight line segments as follows:

$$AB = 1, \quad BC = \phi, \quad CD = \phi^3 = 2\phi + 1 \quad \text{(Fig. 8.1)}$$

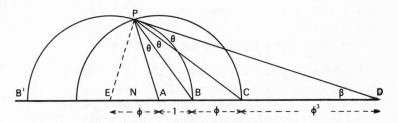

Fig. 8.1. Trisection of an angle

We seek a point P such that $\angle APB = \angle BPC = \angle CPD = \theta$.

PB is the bisector of $\angle APC$; hence, by a well-known theorem, P lies on the locus of points for which $PC:PA = BC:BA = \phi:1$. This locus is a circle having BB' as diameter, where B' divides AC externally in the same ratio as B divides it internally. Then it is easy to show that E is the center of this circle of radius $EB = \phi + 1 = AC$.

Again, since PC is the bisector of $\angle BPD$, P lies on the locus of points for which $PD:PB = CD:CB = \phi^3:\phi = (\phi + 1):1$.

This locus is a circle passing through C. Its radius can be shown to be $\phi + 1$ also, its center being at A, where $AC = \phi + 1$.

These two circles of equal radius intersect above the line $B'D$ at the point P. Thus P is the required point.

By using the dimensions and labels shown in figure 8.1 it is easily seen that $\cos \angle PAE = 1(2\phi)$, so that $\angle PAE = 72°$ (see p. 40).

By expressing CP in terms of ϕ (if PN is perpendicular to ND, $CP^2 = PN^2 + NC^2$) it can be shown to equal CD so that $\beta = \theta$.

Hence $3\theta + \beta = 4\theta = \angle PAE = 72°$. Therefore $\theta = \beta = 18°$.

An alternative method of solution is to consider the cross-ratio: $P\{ABCD\}$

$$\frac{BC \cdot AD}{AB \cdot DC} = \frac{\sin \theta \cdot \sin 3\theta}{\sin \theta \cdot \sin \theta}$$

whence

$$\frac{\phi(\phi^2 + \phi^3)}{\phi^3} = \frac{\sin 3\theta}{\sin \theta} = 4 \cos^2 \theta - 1$$

Hence $4 \cos^2 \theta = \phi^2 + 1 = \phi + 2$. Therefore $\theta = 18°$ (see p. 40).

PHI: ANOTHER HIDING PLACE

The following example produces a useful result. It allows us to find in a speedy and straightforward fashion the position of the golden cut of a line of given length below an upper limit. In addition, it is attractive in its simplicity and provides a further instance of our unexpectedly uncovering a novel hiding place for *Phi*.

A circle B is cut out from a circle A which it touches internally at O (Fig. 8.2). The area of B is such that the centroid of the lunar remnant $(A - B)$ lies on the circumference of B at C.

Let the linear dimensions of A and B be in the ratio $a:b$. Then the areas of the two circles and the lune may be written as λa^2, λb^2, and $\lambda(a^2 - b^2)$. Let C_a, C_b, and C be the centroids of A, B, and the lunar remnant. These are collinear. Taking moments about O,

$$\lambda a^2 \cdot OC_a - \lambda b^2 \cdot OC_b = \lambda(a^2 - b^2) \cdot OC$$

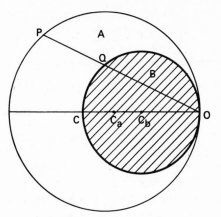

Fig. 8.2. Centroid of a lune

Dividing by $\lambda b^2 \cdot OC_b$

$$\frac{a^2}{b^2} \cdot \frac{OC_a}{OC_b} - 1 = \left(\frac{a^2}{b^2} - 1\right) \cdot \frac{OC}{OC_b}$$

But

$$\frac{OC_a}{OC_b} = \frac{a}{b} \quad \text{and} \quad \frac{OC}{OC_b} = 2$$

Hence

$$\frac{a^3}{b^3} - 1 = 2\left(\frac{a^2}{b^2} - 1\right)$$

Since $(a/b - 1) \neq 0$,

$$\frac{a^2}{b^2} + \frac{a}{b} + 1 = 2\left(\frac{a}{b} + 1\right)$$

i.e.,

$$\frac{a^2}{b^2} - \frac{a}{b} - 1 = 0$$

whence

$$\frac{a}{b} = \phi \text{ or } \phi'$$

It follows that any chord OP of circle A is cut in the golden ratio by the circumference of $B(Q)$.

The example can be generalized to apply to two similar plane figures A and B of *any shape*, regular or irregular, provided that the straight line through C_a, C_b, and the common point O is similarly situated with respect to A and B.

THE TETRAHEDRON PROBLEM

This example is more difficult than the preceding. Nevertheless, when it was published in the mathematical column of the well-known *Journal of the Assistant Masters Association* (Feb. 1966, p. 75), one of the first correct solutions received was from a seventeen-year-old school boy. The problem is included here with the kind permission of the editor as a further instance of the ubiquity of *Phi*. Solvers seemed to enjoy the problem and expressed their thanks, using such epithets as "nice," "amusing" ...! Several pointed out that the enunciation of the problem should have excluded the case of four congruent triangles. This, however, is a trivial solution.

Here is the problem:

The faces of a tetrahedron are all scalene triangles similar to one another, but not all congruent, with integral sides. The longest side does not exceed 50. Show its network.

The limitation to integral values being waived, show that the ratio of the length of the longest to that of the shortest edge has a limiting value, and find it.

The solution is as follows:

Two triangles may have five parts of the one congruent with five parts of the other without being congruent triangles. If the triangles are not congruent, their congruent parts cannot include the three sides. Hence the triangles must be equiangular, and it is easily shown that the lengths of the sides must be in *geometrical* progression: a, λa, $\lambda^2 a$. If a is the shortest edge, $\lambda^3 a$ is the longest ($\lambda > 1$).

Since $a + \lambda a > \lambda^2 a$,

$$\lambda^2 - \lambda - 1 < 0, \text{ or } (\lambda - \phi)(\lambda - \phi') < 0$$

where ϕ, ϕ' are the roots of

$$\lambda^2 - \lambda - 1 = 0 \quad (\phi = (1 + \sqrt{5})/2 = 1.6180\cdots,$$
$$\phi' = -1/\phi = (1 - \sqrt{5})/2 = -0.6180\cdots).$$

Hence

$$\phi' \leqslant \lambda \leqslant \phi$$

The inequality shows that λ cannot be greater than ϕ, the "golden ratio" of the Greeks, so that the limiting value of the ratio of the length of the longest edge to that of the shortest of the tetrahedron is $\phi^3 = \sqrt{5} + 2$.

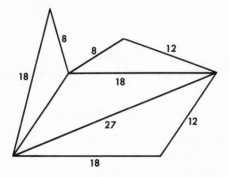

Fig. 8.3. Tetrahedron net

To construct the network of the tetrahedron fulfilling the specification, we require four lengths in geometrical progression: a, λa, $\lambda^2 a$, $\lambda^3 a$. Since they are integral, λ must be a simple ratio such as $3/2$. In this case, $\lambda^2 = 9/4$ and a must be a multiple of 4; $\lambda^3 a < 50$ means that $a < 15$.

The required solution is $a = 8$, $\lambda a = 12$, $\lambda^2 a = 18$, $\lambda^3 a = 27$ as edges of the tetrahedron. Figure 8.3 shows the net.

TWO TRIANGLES

For the following interesting addition to this Anthology I have to thank a correspondent, F. M. Goldner, Esq.

A triangle ABC has sides such that $a > b > c$. A second triangle has sides $1/a$, $1/b$, $1/c$.

Prove that a/c has an upper limit and find it.

$$a - c < b, \qquad \frac{1}{c} - \frac{1}{a} < \frac{1}{b}$$

By multiplication, $(a - c)^2/ac < 1$, i.e., $(a^2/c^2) - 3(a/c) + 1 < 0$, i.e.,

$$\left(\frac{a}{c} - \frac{3 + \sqrt{5}}{2}\right)\left(\frac{a}{c} - \frac{3 - \sqrt{5}}{2}\right) < 0$$

The first factor being smaller than the second, the first must be negative and the second positive; hence

$$\frac{a}{c} < \frac{3 + \sqrt{5}}{2}$$

i.e., $a/c < \phi^2$, the required upper limit.

The next problem brings us from geometry to analysis.

LOG OF THE GOLDEN MEAN[2]

Find the value of a satisfying the equation $n^n + (n + a)^n = (n + 2a)^n$ in the limit as $n \to \infty$.

We use the fact that $\lim\limits_{n \to \infty} [1 - (x/n)]^n = e^x$.

Dividing the given equation by n^n, we have

$$1 - \left(1 - \frac{a}{n}\right)^n = \left(1 - \frac{2a}{n}\right)^n$$

In the limit as $n \to \infty$ this is $1 + e^a = e^{2a}$.

Writing $e^a = y$, $y^2 - y - 1 = 0$. Thus $y = e^a = \phi$. Therefore $a = \log_e \phi$. Thus, a is the natural logarithm of the golden mean:

$$a = \log_e \frac{1 + \sqrt{5}}{2} \qquad \text{or} \qquad e^a = \phi$$

◻ ◻ ◻

The concluding pages of this chapter will serve as interesting reading as well as useful practice for those who have an acquaintance with the coordinate geometry of the conic sections. If the

eccentricities of these curves are expressed in terms of the golden
section, the curves are linked to each other in a truly fascinating
manner. Only a few of the remarkable results of connecting e and ϕ
are listed here, but the reader may be glad to find his own further
evidence of harmony. These results, which (as far as the writer is
aware) have not been published before, stand as a demonstration
of the fact that even in the field of elementary mathematics, some
jewels remain to be uncovered.

PHI AND THE PARABOLA

Interesting properties are associated with the focal chord PQ
of the parabola $y^2 = 4ax$, when the parameter of P is the golden
section, ϕ, i.e., P is $(a\phi^2, 2a\phi)$.

Referring to figure 8.4, let the focal chord PSQ be produced to
cut the y-axis in A, the directrix in R. Let LL' be the latus rectum,
T be $(6a, 0)$ and $\angle PST = \theta$.

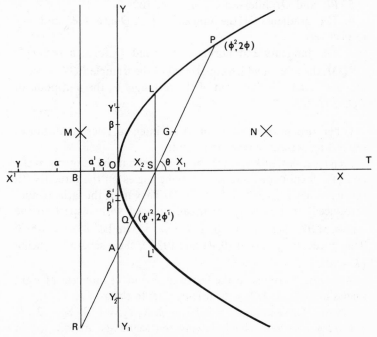

Fig. 8.4. *Phi* and the parabola $y^2 = 4x$

Let *PL* produced meet the *x*- and *y*-axes in α, β; $L'Q$ meet these axes in α', β'; PL' in X_1, Y_1; QL in X_2, Y_2.

Let the tangents at *P*, *Q* meet the *x*- and *y*-axes in γ, γ', and δ, δ'.

Properties of this focal chord are interesting on account of their simplicity and the number of segments expressible in integers. To bring this out the more clearly we let $a = 1$ in $y^2 = 4ax$. The interested reader will derive some satisfaction from verifying the following.

1. The focal chord from $P(\phi^2, 2\phi)$ terminates in *Q*, and *Q* is $(\phi'^2, 2\phi')$. It intersects the *y*-axis and the directrix in $(0, -2)$ and $(-1, -4)$ respectively.

2. *S* being the focus, $PS = \phi + 2$, $QS = \phi' + 2$, so the length of the focal chord is 5. Its gradient is 2, i.e., $\tan \theta = 2$.

3. *T* being $(6, 0)$, *P*, *O*, *Q*, *T*, and *L'* are concyclic.

4. $PR = 5\phi$, $QR = -5\phi'$.

5. *PL'* and *QL* intersect on the directrix.

6. The gradients of the tangents at *P*, *Q* are $-\phi'$ and $-\phi$ respectively.

7. The tangents and normals at *P* and *Q* form a rectangle *PNQM*, the area of which equals that of the triangle $PQT = 5\sqrt{5}$, i.e., $5(\phi - \phi')$. The centroid of the rectangle *G*, the mid-point of *PQ*, is $(3/2, 1)$.

Other results expressible in small integers refer to distances separating certain intersection points.

Tangents. A well-known theorem states that the tangents at the end of a focal chord intersect at right angles on the directrix. In this case they intersect at *M*, $(-1, 1)$. They meet the latus rectum (produced) in points symmetrically placed with respect to the *x*-axis, at distances from it of $\phi - \phi' = \sqrt{5}$ and $\phi' - \phi = -\sqrt{5}$. They meet the *y*-axis at $(0, \phi)$ and $(0, \phi')$, their separation being $(\phi - \phi')$, i.e., $\gamma'\delta' = \sqrt{5}$.

Normals. Normals to the curve at *P* and *Q* intersect at right angles at $N(4, 1)$. $MN = 5$; it is parallel to the *x*-axis.

Chords. These are: $\alpha\alpha' = 1$, $\beta\beta' = 2$, $X_1X_2 = 1$, $Y_1Y_2 = 2$.

Golden Sections. The following ten segments are divided in the golden ratio $(\phi : 1)$ by the points indicated:

	SEGMENT	POINT		SEGMENT	POINT
1	PSA	S	6	$L\beta'\alpha'$	β'
2	SQA	Q	7	$PL\beta$	L
3	αLP	L	8	$L'Q'\beta'$	Q
4	$\alpha'QL'$	Q	9	PX_1L'	X_1
5	$L\beta\alpha$	β	10	LX_2Q	X_2

Finally, if the parabola is drawn on the ellipse/hyperbola figure (Fig. 8.5) with the same origin and on the same scale, then:

1. The latus rectum of the parabola is the directrix of the hyperbola.

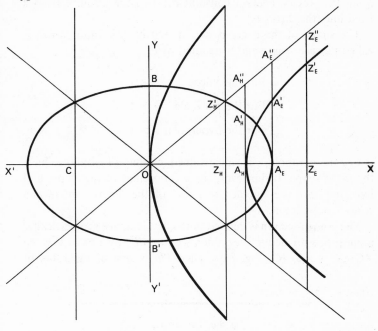

Fig. 8.5. Golden and conic sections

ELLIPSE	PARABOLA	HYPERBOLA	ASYMPTOTES
$\dfrac{x^2}{\phi^2} + \dfrac{y^2}{1} = 1$	$y^2 = 4x$	$\dfrac{x^2}{\phi} - \dfrac{y^2}{1} = 1$	$y = \pm\dfrac{b}{x} = \pm\dfrac{x}{\sqrt{\phi}}$
$e^2 = -\phi'$	$e^2 = 1$	$e^2 = \phi$	
$a = \pm\phi$		$a = \pm\sqrt{\phi}$	
$b = 1$		$b = 1$	

2. The directrix of the parabola is the image in the y-axis of the directrix of hyperbola.

3. The hyperbola asymptotes intersect the parabola in the points $(4\phi, 4\sqrt{\phi})$ and $(-4\phi, -4\sqrt{\phi})$.

GOLDEN AND CONIC SECTIONS

A conic section—ellipse, hyperbola or parabola—may be considered either as a section of a right circular cone by a plane, or as the locus of a point which moves so that its distance from a fixed point (the focus) bears a constant ratio to the distance from a fixed line (the directrix).

The shape of the curve is determined by this ratio, which is called its *eccentricity* and is denoted by e.

$$\text{For the ellipse,} \quad e < 1$$

$$\text{For the parabola,} \quad e = 1$$

$$\text{For the hyperbola,} \; e > 1$$

For the parabola, e is single-valued, so that all parabolae have the same shape. If, however, the eccentricities of the ellipse and hyperbola are the "golden section" of the Greeks, interesting relationships result.

The golden section is defined in the first instance geometrically: a line AB is divided in the golden section, *Phi*, by a point C when $AC/CB = AB/AC$ (Fig. 8.6). Then *Phi* is one of the roots of

Fig. 8.6. Golden cut

$x^2 - x - 1 = 0$. As we know, $\phi = (1 + \sqrt{5})/2 = 1.6180\cdots$, the other root being $\phi' = (1 - \sqrt{5})/2 = -0.6180\cdots$.

$$\phi\phi' = -1, \quad \phi + \phi' = 1, \quad \phi - \phi' = \sqrt{5}$$

If the eccentricities of the conic sections are given the following values, the major and minor axes of the ellipse and hyperbola are determined:

Ellipse: $e^2 = -\phi'$, $b^2 = a^2(1 - e^2) = a^2(\phi' + 1) = a^2\phi'^2$

Parabola: $e^2 = (\phi + \phi')^2$

Hyperbola: $e^2 = \phi$, $b^2 = a^2(e^2 - 1) = a^2(\phi - 1) = -a^2\phi'$

Let us, for simplicity, ascribe unit length to b. Then

For the ellipse: $e = \sqrt{-\phi'}, a = \pm\phi, b = 1, \dfrac{x^2}{\phi^2} + \dfrac{y^2}{1} = 1$

For the hyperbola: $e = \sqrt{\phi}, a = \sqrt{\phi}, b = 1, \dfrac{x^2}{\phi} - \dfrac{y^2}{1} = 1$

Setting out these two curves and the asymptotes of the hyperbola $y = \pm x/\sqrt{\phi}$ on coordinate axes (see Fig. 8.5), so that their centers coincide with the origin, the significant points and lines associated with them being marked, we observe the following facts concerning the distances of points on the x-axis from the origin O:

a. they are all simple functions of ϕ
b. they form a geometrical progression
c. there are remarkable coincidences (at A_H and A_E).

POINT	DISTANCE FROM O	DESCRIPTION
O	0	Origin
Z_H	ϕ^0	Hyperbola directrix
A_H	$\phi^{1/2}$	{Ellipse focus / Hyperbola vertex
A_E	ϕ	{Ellipse vertex / Hyperbola focus
Z_E	$\phi^{3/2}$	Ellipse directrix

Certain segments are divided in the golden ratio in a reciprocal fashion:

$$OA_E \quad \text{by} \quad Z_H$$
$$OZ_E \quad \text{by} \quad A_H$$

Perpendiculars to the axis at the five points being cut by the ellipse, hyperbola, and asymptote, the lengths of all the segments may be expressed as functions of ϕ, and more coincidences are revealed:

SEGMENT	LENGTH	DESCRIPTION
OB	$\phi + \phi' = 1$	Ellipse: semi-minor axis / Hyperbola: semi-minor axis
$Z_H Z'_H$	$\sqrt{-\phi'}$	Intercept of hyperbola directrix by ellipse / Intercept of hyperbola directrix by its asymptote
$A_H A'_H$	$-\phi'$	Ellipse: semi-latus rectum / Hyperbola: tangent at vertex
$A_H A''_H$	$\phi + \phi' = 1$	Ellipse: semi-latus rectum produced / Hyperbola: tangent at vertex
$A_E A'_E$	$\sqrt{-\phi'}$	Ellipse: tangent at vertex / Hyperbola: semi-latus rectum
$A_E A''_E$	$\sqrt{\phi}$	Intercept of tangent at ellipse vertex and of hyperbola semi-latus rectum produced, by asymptote
$Z_E Z'_E$	$\sqrt{\phi}$	Intercept of ellipse directrix by hyperbola
$Z_E Z''_E$	ϕ	Intercept of ellipse directrix by asymptote

Note especially:

1. The concurrence at Z'_H of ellipse, hyperbola directrix and asymptote.

2. The reciprocation between $A_H A'_H$ and $A_E A'_E$.

3. The reciprocation between ordinates at Z_H and Z_E.

4. The following equalities:

$$\text{i. } OB = A_H A''_H = \phi + \phi'$$
$$\text{ii. } Z_H Z'_H = A_E A'_E = \sqrt{-\phi'}$$
$$\text{iii. } A_E A''_E = Z_E Z'_E = \sqrt{\phi}$$

5. Certain segments are divided in the golden ratio:

$$\text{i. } A_H A''_H \quad \text{by} \quad A'_H$$
$$\text{ii. } A_E A''_E \quad \text{by} \quad A'_E$$

The following details relative to the *ellipse* are worthy of note:

Any transversal of the y-axis and the directrix (Z_E) is divided in the golden ratio by the latus rectum of the ellipse. In particular,

1. The focus A_H so divides OZ_E.
2. A''_H so divides OZ''_E.
3. The tangent to any ellipse at the end point of its latus rectum (A'_H) intersects the directrix at the point where it crosses the major axis produced (Z_E). In this case the point of tangency (A'_H) divides the transversal between the axes *internally* in the golden ratio. Moreover, the lengths of the intercepts on the x- and y-axes are $\phi^{3/2}$ and ϕ (=the semi-major axis) respectively.
4. The area of the ellipse is $\pi\phi$.

The following details relative to the *hyperbola* are parallel to these:

Any transversal of the y-axis and the latus rectum (A_E) is divided in the golden ratio by the directrix (Z_H). In particular:

1. The point Z_H so divides OA_E.
2. The point Z'_H so divides OA''_E.
3. The tangent to any hyperbola at the end point of its latus rectum (A'_E) intersects the directrix at the point where it crosses the transverse axis (Z_H). In this case the point of tangency (A'_E) divides the transversal between the axes *externally* in the golden ratio. Moreover, the lengths of the intercepts on the x- and y-axes are $\phi + \phi'$ and $\sqrt{\phi}$ (= the semi-transverse axis) respectively.

Finally, referring to both curves, the corresponding end-points of their latera recta are collinear with the origin of coordinates, O.

Patterns

It is a matter of common observation that patterns can be a source of aesthetic pleasure, whether they are found in nature or in the creative output of a mathematical imagination. A snowflake is a pattern comprised of equilateral triangles of identical design joined to form a hexagon. The honeycomb is comprised of conjoined hexagons. In chapter V, figure 5.5, we found that a pattern of golden rectangles gave rise to an equiangular spiral. Other examples will occur to the reader.

The mental activity involved in the appreciation of a pattern is that of perceiving relationships. That this is an ancient activity is shown by the way in which ancient unremembered astronomers, before the Christian era, brought certain bright stars into association with one another to form the constellations: Orion, the Hunter; Delphinus, the Dolphin; Aquila, the Eagle; Ursa Major, the Great Bear, and others.

When one contemplates a pattern, its various parts must be mentally related to the whole and the pattern must be grasped and appreciated as a whole. The pattern of notes of a melody form a sequence in time, but unless memory allows the whole to be grasped in an instant, the beauty vanishes. The repetitive contiguous hexagons of the honeycomb are seen as a unity; likewise the expanding equiangular spiral segments of the nautilus sea shell. Their beauty lies in "the reduction of the many to the one."

As well as visual patterns and designs, there are aural examples. Rhythm is a pattern in sound. It can be simple, like the pulse beat, or it can be complex, like some Indian drum music. Rhythm is heard in poetry as well as in music. Sometimes both are conspicuous simultaneously as in Coleridge Taylor's musical setting of Longfellow's "Hiawatha."

CHESS PROBLEMS

An example of a design superimposed on a background pattern, conceived and constructed to provide aesthetic pleasure, is a problem set out on a chess board. Figure 9.1, in which White is to

BLACK

WHITE

Fig. 9.1. White to play and mate in two

play and mate in two moves, is an illustration. A chess problem, Hardy writes, is "simply an exercise in pure mathematics...," and "...chess problems are the hymn-tunes of mathematics."[1]

Hogben's remark that "the aesthetic appeal of mathematics may be very real for the chosen few" implies that the discipline has no widespread appeal. To realize that this is a fallacy, one has only to think of the multitudes of chess enthusiasts throughout the world. Supporting evidence is found in the wide variety of newspapers and other journals in many languages which regularly publish chess problems. Few will question that the composing or the solving of a chess problem is a rich source of aesthetic pleasure. An eloquent witness to this is Vladimir Nabokov:

> Frequently, in the friendly middle of the day, on the fringe of some trivial occupation, in the idle wake of a passing thought, I would experience, without warning, a spasm of acute mental pleasure, as the bud of a chess problem burst open in my brain, promising me a night of labour and felicity. It might be a new way of blending an unusual strategic device with an unusual line of defence; it might be a glimpse, curiously stylised and thus incomplete, of the actual configuration that would render at last, with humour and grace, a difficult theme that I had despaired of expressing before.... Whatever it was, it belonged to an especially exhilarating order of sensation.... The strain on the mind is formidable; the element of time drops out of one's consciousness altogether.
>
> But whatever I can say about this matter of problem composing I do not seem to convey sufficiently the ecstatic core of the process and its points of connection with various other, more overt and more fruitful, operations of the creative mind.... The event is accompanied by a mellow physical satisfaction, especially when the chessmen are beginning to enact adequately, in a penultimate rehearsal, the composer's dream....[2]

EULER'S FORMULA

Another example of the satisfaction to be found in patterns and designs in mathematics is a famous one that interrelates convex polyhedra, including the quintet of five regular polyhedra familiar to the ancient Greeks. Mystical properties were ascribed to these, but in more recent times a simple and beautiful relationship between them has been revealed by the genius of one of the greatest mathematicians: Euler.

Euler's formula is easily stated. If F, V, and E denote the number of faces, vertices, and edges of any convex polyhedron, then

$$F + V = E + 2$$

The truth of this relationship in respect of the five polyhedra mentioned above may be tested at once by reference to the following table:

	F	V	E
Tetrahedron	4	4	6
Cube	6	8	12
Octahedron	8	6	12
Icosahedron	20	12	30
Dodecahedron	12	20	30

But the formula applies to any convex polyhedron. To confirm this, take as an example a square pyramid mounted on a cube (Fig. 9.2). Here $F = 9$, $V = 9$, $E = 16$.

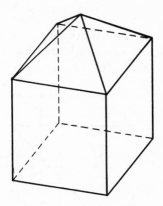

Fig. 9.2. Pyramid on a cube

Another example might be a truncated pentagonal pyramid (Fig. 9.3) in which $F = 7$, $V = 10$, $E = 15$.

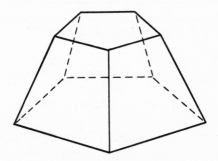

Fig. 9.3. Truncated pyramid

If the reader will make a few similar experiments for himself, he will find that the verification of Euler's formula affords him some mental satisfaction. He will then be in a position to inquire, introspectively, wherein the pleasure lies. It may prove that the recipe for such pleasure includes such ingredients as the neatness and brevity of Euler's formula, flavored with the spice of surprise and the sense of power which stems from bringing all convex polyhedra under one simple, all-embracing rule.

MAGIC SQUARES

An effective if somewhat trivial example of an arithmetic pattern is the magic square, in which each row of numbers, each column and each diagonal adds to the same number (Fig. 9.4). The cult of

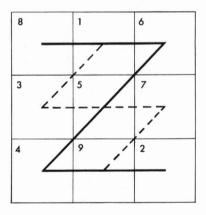

Fig. 9.4. Magic square (order 3)

the magic square has existed for many centuries and numerous books have been published on the subject.

The *order* of a magic square is the number of "cells" which make up one side of the square. There is only one square of order three but there are 880 of order four. The number of magic squares of order five is not known, but it is in the millions.

Although a section on magic squares may seem to be too

frivolous to be worthy of consideration under the heading of beauty in mathematics, we shall see that it justifies itself as an example of patterns and designs. And as to the general topic of this book, a type of magic square can, as we shall see, be made from members of the Fibonacci series.

The only third-order magic square is shown in figure 9.4. By joining the odd-numbered cells in serial order with a line, and the even-numbered cells similarly with a line, we obtain the simple pattern shown in the figure.

The fourth-order squares contain a species with very remarkable properties. These are called "diabolic." An example is shown in

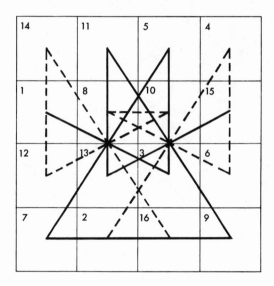

Fig. 9.5. Magic square (order 4)

figure 9.5. Not only does this exhibit the usual properties—rows, columns, and diagonals adding to the same total (34)—but several other groups of 4 cells—the four corners cells, any group of 2 by 2 cells, etc.—also add to 34. Again, by joining the odd-numbered cells in serial order with a line and the even-numbered cells serially with another line, we obtain the type of symmetry shown in figure 9.5.

The interchange of the top and bottom rows, or of the left-hand and right-hand columns, or both, does not change the properties of the square, but it will of course change the appearance of the line patterns, all of which show symmetry about a square bisector. A more complicated pattern is obtained by joining the numbered cells in serial order without distinguishing between odd and even

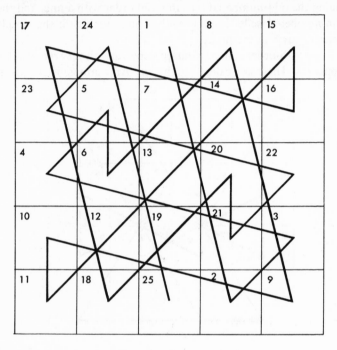

Fig. 9.6. Magic square (order 5)

numbers. A pattern of this type is shown for magic square of order five (65) in figure 9.6.

The unfailing emergence of these patterns in every magic square is unexpected and intriguing. Why should any and every magic square of any order give rise to such attractive patterns, no two alike?

It is a simple matter to form a magic square of any order in which the *products* of the rows, columns and diagonals are the same. An example of the third order follows:

128	1	32
4	16	64
8	256	2

In this the product is 4096 and each cell contains a power of 2 from $2^0 = 1$ to $2^8 = 256$, the indices being the numbers found in figure 9.4.

A sort of magic square can be formed with any consecutive members of the Fibonacci series. An example will make this clear.

Take *any* 9 consecutive numbers from the following series:

$$0 \quad 1 \quad 1 \quad 2 \left| \begin{array}{ccccccccc} v_1 & v_2 & v_3 & v_4 & v_5 & v_6 & v_7 & v_8 & v_9 \\ 3 & 5 & 8 & 13 & 21 & 34 & 55 & 89 & 144 \end{array} \right| 233 \quad 377 \quad \cdots$$

Now form a magic square in which the subscripts of v add to 15 (Fig. 9.4).

v_8	v_1	v_6
v_3	v_5	v_7
v_4	v_9	v_2

Substitute for these the corresponding Fibonacci numbers:

89	3	34
8	21	55
13	144	5

The sum of the products of the three *rows* is 9078 + 9240 + 9360 = 27678. The sum of the products of the three *columns* is 9256 + 9072 + 9350 = 27678.

POLYGONAL NUMBERS

Another byway in which one finds interesting examples of mathematical patterns is that of polygonal numbers. We shall consider triangular, square, and especially—in harmony with the theme of our anthology—*pentagonal numbers*. We shall again discover curious and surprising relationships which link these number series, but their *rationale* is easier to grasp than that of the mysterious patterns issuing from magic squares.

The subject is a much larger one than the brevity of this notice might suggest. Enough is said, however, to fulfil the present purpose, which is to provide clear and compelling examples of the beauty and fascination of some of the patterns which illuminate much of the territory of elementary mathematics.

We have seen that the disciples of Pythagoras were especially interested in the regular five-sided polygon, which has been shown to harbor examples of the golden section. Their order had chosen for its symbol the interwoven triple triangle—the pentagram (p. 28). The relation of the golden section to various geometrical ratios of this symbol is set out on p. 29. We have seen that the *sectio auri* is associated also with the Fibonacci series, with the roots of $x^2 - x - 1 = 0$, and with other byways of mathematics.

We now make an excursion into a topic which well deserves a place under the heading of "patterns"; it finds a connection with our general theme through *pentagonal numbers*. That the subject cannot be regarded as trivial is indicated by the circumstance that —as we shall see—it was not beneath the notice of some of the greatest mathematicians, such as Fermat and Euler.

Polygonal numbers are series whose name derives from their association with the shapes of regular polygons—triangle, square, pentagon, etc. The formation of such a series may be illustrated by considering the *triangular numbers*, which are diagrammatically represented by the following patterns:

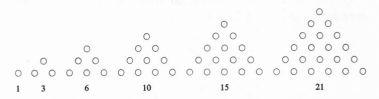

The number of points in each such triangular formation is evidently the sum of an arithmetical progression. By counting the number of points in each row, starting from the top of the triangle, we have $1 + 2 + 3 + 4 + 5 + \cdots$, the common difference being 1.

The triangular number series has some interesting properties, of which two may be mentioned here:

1. The nth triangular number being $n(n + 1)/2$ and the $(n + 1)$th being $(n + 1)(n + 2)/2$, their sum is

$$\frac{n(n + 1)}{2} + \frac{(n + 1)(n + 2)}{2} = \frac{n + 1}{2}(n + n + 2) = (n + 1)^2$$

Thus the sum of two consecutive triangular numbers is a perfect square; the square number series is related to the triangular number series.

2. If unity be added to eight times a triangular number we obtain the square of an odd number:

$$\left(\frac{n(n + 1)}{2} \times 8\right) + 1 = 4n^2 + 4n + 1 = (2n + 1)^2$$

The square numbers may be displayed after the manner of the triangular numbers:

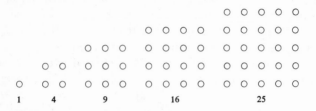

Evidently, the nth square number is n^2.

Coming now to the pentagonal numbers, the first five are represented in figure 9.7.

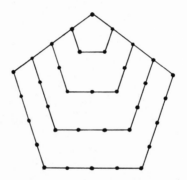

Fig. 9.7. Pentagonal numbers

The first twelve numbers of the pentagonal number series are:

1   5   12   22   35   51   70   92   117   145   176   210

We have seen that the triangular number series is such that any member of it is the sum of an arithmetical progression having a common difference of unity. Consider now the arithmetical progression having a common difference equal to 3:

1   4   7   10   13   16   19   22   25 $\cdots$

We have selected every third member of the natural number series. If we now sum the first n numbers of this series, we obtain the nth member of the pentagonal number series. For example, the fifth member of this series is

$$1 + 4 + 7 + 10 + 13 = 35$$

Since the nth member of the above arithmetical progression is $3n - 2$, the sum of the first n terms is

$$[1 + (3n - 2)] \cdot \frac{n}{2} = \frac{3n^2 - n}{2}$$

This turns out to be the nth member of the pentagonal number series. To illustrate, take $n = 10$. The sum of the first ten terms of the arithmetical progression is $[3(10)^2 - 10]/2 = 145$, which is the tenth pentagonal number.

We may generalize the foregoing examples of polygonal numbers with a formula that applies to a polygon of any number of sides. The polygonal numbers series are infinite series but we may determine any member of the series by simple substitutions.

Suppose that the polygon has r sides, and denote the nth member of this polygonal series by the symbol P_r^n. Then—

$$P_r^n = \frac{n}{2}[2 + (n - 1)(r - 2)] = n + \frac{n(n - 1)}{2}(r - 2)$$

The proof of this formula is omitted but we may test it against results already obtained.

Thus, for the triangular number series $r = 3$, so that the nth member of the series is $P_3^n = n + n(n - 1)/2 = n(n + 1)/2$, as already found.

The tenth member of the pentagonal number series was found to be 145. Writing $n = 10$, $r = 5$ in the formula, we obtain

$$P_5^{10} = 10 + \frac{10(10 - 1)}{2}(5 - 2) = 145$$

Since $P_5^n = \frac{1}{2}(3n^2 - n) \approx \frac{3}{2}n^2$ for large values of n, it follows that two-thirds of each pentagonal number, for sufficiently large n, approximates to a member of the square number series.

FERMAT'S RULES

We have already mentioned that the polygonal number series are not so trivial as they might seem to be at first acquaintance.

The following statements will confirm this. They are due to Pierre de Fermat (1601–1665), who was actually a lawyer by profession and probably the world's greatest "amateur mathematician." He found that a remarkably simple relationship exists between the polygonal number series and all the natural numbers. This may be expressed in a series of statements as follows:

1. Every number is either triangular, or the sum of *two*, or *three* triangular numbers.

2. Every number is either square, or the sum of *two*, *three*, or *four* square numbers.

3. Every number is either pentagonal, or the sum of *two*, *three*, *four*, or *five* pentagonal numbers.

4. Every number is either hexagonal, or the sum of *two*, *three*, *four*, *five*, or *six* hexagonal numbers.

5. And so on.

We turn next to a pattern which exhibits the twin virtues of beauty and utility. It is sufficiently remarkable and important to justify a new chapter.

Pascal's Triangle and Fibonacci

Pascal's triangle is perhaps the most famous of all number patterns. It is of very ancient standing, being probably more than a thousand years old. Its hidden properties have been revealed more and more as mathematics has developed through the centuries. It is certain that the pattern was known to the Chinese in the thirteenth century, and Tartaglia, born at Brescia in 1500, used it to find the coefficients of x in the expansion of $(1 + x)^n$ for a limited number of cases. Pascal made fuller use of the triangle in his *Traité du Triangle Arithmétique*, which he wrote around 1653. There he uses it both to find the coefficients in the binomial expansion $(a + b)^n$ and also to solve combinatorial problems: how many combinations are there of n different things taken r at a time? Pascal also used the triangle to solve problems in probability.

It is evident that a pattern like the arithmetical triangle must be included in the bracket of serious mathematics, as distinct from mathematics which, while serving as a setting for gems like chess problems and magic squares, have borne fruit neither in important applications nor as instruments of further exploration of the unknown. There are, of course, a very large number of mathematical puzzles and games, which, incidentally, testify to the widespread popularity of the mathematical discipline; many of these can stimulate aesthetic pleasure. But, unless they reveal some value as tools or open a door to fields of fresh mathematical enquiry, we cannot regard them as *serious* mathematics, which, as in the case of the present section, is both beautiful in itself and valuable in its many-sided usefulness.

Despite the variety of its applications, Pascal's triangle is of very simple construction. At the apex of the triangle is the figure 1 (row $n = 0$). The row $n = 1$ consists of two figures, 1 and 1. All other figures are the sums of the two just above them, e.g., $10 = 4 + 6$.

n					a					
0				b	1					
1			c	1		1				
2		d	1		2		1			
3	e	1		3		3		1		
4	f	1	4		6		4		1	
5	g	1	5	10		10		5	1	
6	h	1	6	15	20		15	6	1	
7	i	1	7	21	35	35	21	7	1	
8	j	1	8	28	56	70	56	28	8	1
9	1	9	36	84	126	126	84	36	9	1
10	1	10	45	120	210	252	210	120	45	10	1

Certain other numerical relations are to be noted:

a. The sum of the numbers in the nth row is 2^n.

b. The sum of all the numbers above the nth row is $2^n - 1$.

c. The sum of the numbers in any diagonal $(a, b, c, \cdots)$ is the number south-west of the last number included.

Example. Diagonal d: $1 + 4 + 10 + \cdots + 84 = 210$.

d. Any row numbered $2^n - 1$ consists of odd numbers only.

Example. Let $n = 3$. The numbers in row 7 are all odd.

e. Consider a definite intersection point, such as the intersection of row 6, diagonal c (number 15). How many different routes, beginning at the apex, may be taken by a point which moves from intersection to intersection always in a s.e. or s.w. direction? The answer is 15. The number 210 may be reached from the apex in this way by precisely 15 different routes.

f. If n is a prime number, n is a factor of all the numbers in the nth row except the 1's at the ends.

One of the first uses to be made of Pascal's triangle was to find the coefficients of the expansion of $(1 + x)^n$. These are found in the nth row.

Example: $(1 + x)^3 = 1 + 3x + 3x^2 + x^3$ (Row 3).

CHINESE TRIANGLE

Properties like these are sometimes made more evident by a modification of the array. By displacing the numbers of Pascal's triangle to the left, as shown in the following array, we arrive at the "Chinese triangle," as it was known to Fibonacci.

r		0	1	2	3	4	5	6	7	8	9	10	11	12	⋯
FIBONACCI SERIES		0	1	1	2	3	5	8	13	21	34	55	89	144	⋯
n															
0	1														
1	1	1													
2	1	2	1												
3	1	3	3	1											
4	1	4	6	4	1										
5	1	5	10	10	5	1									
6	1	6	15	20	15	6	1								
7	1	7	21	35	35	21	7	1							
8	1	8	28	56	70	56	28	8	1						
9	1	9	36	84	126	126	84	36	9	1					
10	1	10	45	120	210	⋯					1				
11	1	11	⋯											55	
12	1	12	⋯												

From this array we may read off at once the coefficient of the general term of the expansion of $(1 + x)^n$. The coefficient of the $(r + 1)$th term is ^nC_r. The diagonals of Pascal's triangle have become the *columns* of the Chinese triangle.

Example. The coefficient of the 5th term of $(1 + x)^9$ is $^9C_4$. It is found in row $n = 9$ and column $r = 4$, *viz.*, 126.

Example. Column $r = 2$ gives the coefficients of $(1 - x)^{-3} = 1 + 3x + 6x^2 + 10x^3 + \cdots$.

If $+x$ takes the place of $-x$ then the signs alternate:

Example. $(1 + x)^{-2} = 1 - 2x + 3x^2 - 4x^3 + \cdots$.

The column $r = 2$ contains the triangular numbers for two-dimensional space: $1, 3, 6, \cdots$ (p. 127). The column $r = 3$ con-contains the triangular numbers for three-dimensional space: $1, 4, 10 \cdots$. A comparison of the two columns yields interesting results. Twenty tennis balls which can be arranged in a tetrahedal array suffice to form precisely two triangles in a plane. Two triangles each of 28 balls suffice exactly to form one tetrahedral pyramid.

Another feature of Pascal's triangle is that it contains the Fibonacci series, though there seems to be no record to show that Pascal had noticed this. It is possible that Leonardo Fibonacci may have stumbled on the series known today by his name through an examination of the Chinese triangle. It will be seen that if in this latter array numbers are added diagonally—whether the diagonals slope upwards to the right or downwards to the right—the sums reproduce the Fibonacci sequence (see the dashed lines in the array).

We deduce from this that another method of obtaining a member of the Fibonacci series is by a summation of the following type:

$$^nC_0 + {}^{n-1}C_1 + {}^{n-2}C_2 + {}^{n-3}C_3 + \cdots + {}^{n-r}C_r$$

that is,

$$\sum_{r=0}^{n} {}^{n-r}C_r$$

the series terminating when $n - r = 1$ or 0.

Example. Let $n = 9$.

$$^9C_0 + {}^8C_1 + {}^7C_2 + {}^6C_3 = 1 + 8 + 21 + 20 + 5 = 55 = u_{11}$$

That a Fibonacci number can be obtained by a summation along a line in another direction is due to the fact that $^nC_r = {}^nC_{n-r}$:

$$^9C_0 + {}^8C_1 + {}^7C_2 + {}^6C_3 + {}^5C_4 =$$
$$^9C_9 + {}^8C_7 + {}^7C_5 + {}^6C_3 + {}^5C_1 = u_{11}$$

The possibilities of this remarkable arithmetical pattern are not exhausted by this. The Pascal triangle may be used as a calculating device to solve problems in "permutations and combinations." A couple of examples will make this clear.

> *Example.* I wanted to adopt two puppies from the five offered to me. In how many different ways can I select my pair? *Answer.* $^5C_2$.

The answer is found in row $n = 5$, column $r = 2$ in the Chinese triangle: 10.

> *Example.* A hockey club of seven men and nine women has to choose a mixed team of five men and six women. How many different teams are possible? *Answer.* $^7C_5 \times {}^9C_6 = 1764$.

Row $n = 7$, column $r = 5$ gives 21. Row $n = 9$, column $r = 6$ gives 84, and $21 \times 84 = 1764$.

PROBABILITY COMPUTATION

Pascal's triangle can be used as a calculator to determine probabilities. This branch of mathematics, which was at first applied to games, is now of great practical importance. It is for instance a basic tool used by life insurance companies.

We shall illustrate this with simple examples concerned with coin-tossing.

1. If three pennies are tossed, there are just 8 possible results. Representing "heads" by h and "tails" by t,

hhh	*hht*	*tth*	*ttt*
hth	*tht*		
thh	*htt*		
1	3	3	1

These are the numbers in row 3 of the Pascal triangle. Since there are $2^3 = 8$ possibilities, the probability of getting 3 heads (or 3 tails) is 1/8. The probability of getting 2 heads (or 2 tails) is 3/8.

2. In tossing four pennies, what are the chances of getting no heads, one head, two heads, three heads, or four heads? The answers are in row 4 of Pascal's triangle: 1, 4, 6, 4, or 1 divided by 2^4 (the sum of these digits) respectively.

3. If ten coins are tossed, what is the probability of getting five heads and five tails?

With one coin there are two possible different results, with two coins 2×2 results and with ten, 2^{10} different results. *Any* five coins out of the ten showing heads counts as one and five can be selected from ten in $^{10}C_5$ different ways.

Row 10, column 5 of Pascal's triangle gives 252. Thus the required probability is $252/2^{10} = 63/256$, that is, approximately one chance in four.

MORE PASCAL PATTERNS

The features of this elegant number pattern which we have briefly described are sufficient to show that aesthetic pleasure is to be derived from arithmetical patterns, but we have only touched

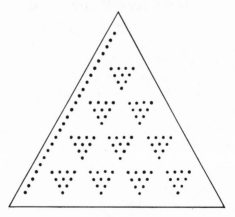

Fig. 10.1. Pascal's triangle: multiples of 5

the fringe of the possibilities. Pascal himself remarked in his treatise on the subject how fertile in properties this triangle proved to be.

The reader may care to search for himself. Can the coefficients of a trinomial expansion $(a + b + c)^m$ be represented in a pattern?

Before we take leave of this topic we would draw attention to the patterns based on Pascal's triangle shown in figures 10.1 and 10.2. Figure 10.1 shows the relative positions in Pascal's triangle

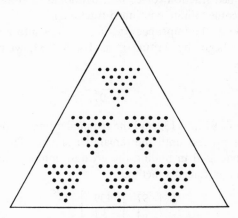

Fig. 10.2. Pascal's triangle: multiples of 7

of all the multiples of 5. In figure 10.2, multiples of 7 are picked out. The results are as unexpected as in the case of the magic squares.

CONTINUED FRACTIONS

As a concluding contribution to this chapter, we exhibit the pattern of the simplest of the continued fractions. Here we meet one of the most interesting appearances of the golden ratio, remarkable for its close association with the "golden numbers," more formally styled the Fibonacci series.

To appreciate the beauty and simplicity of this method of generating the series we make a short diversion into a subject which, because it is often excluded from the mathematics syllabus, may require a few paragraphs of preliminary instruction.

Continued fractions have their own intrinsic interest and surprises. Their pattern-like appearance, e.g.,

$$\frac{157}{225} = \frac{1}{1+} \; \frac{1}{2+} \; \frac{1}{3+} \; \frac{1}{4+} \; \frac{1}{5}$$

justifies their inclusion with the other patterns already discussed.

We begin by showing how a fraction may be expressed as a finite continued fraction (of course, irrational or transcendental numbers become infinite continued fractions).

To transform the improper fraction 236/139 into a continued fraction, we begin by writing it as $1 + 97/139$, which can be expressed as

$$1 + \frac{1}{139/97}$$

We then divide 97 into 139 and the remainder into the divisor and continue the process until the remainder is zero. The lay-out is similar to that used in finding the greatest common measure of two numbers by Euclid's method:

2	97	139	1
	84	97	
4	13	42	3
	12	39	
	1	3	3
		3	
		0	

The required continued fraction is derived from the successive divisors:

$$\frac{236}{139} = 1 + \frac{97}{139} = 1 + \cfrac{1}{1 + \cfrac{1}{2 + \cfrac{1}{3 + \cfrac{1}{4 + \cfrac{1}{3}}}}}$$

This is conventionally written thus:

$$\frac{236}{139} = 1 + \frac{1}{1+} \frac{1}{2+} \frac{1}{3+} \frac{1}{4+} \frac{1}{3}$$

If we ignore the last three divisors and consider only the fraction $1/(1 + \frac{1}{2})$, we obtain 2/3 as a first approximation to 97/139. This is called the *first convergent*.

The second convergent is

$$\cfrac{1}{1+ \cfrac{1}{2+ \cfrac{1}{3}}}$$

which is 7/10, a closer approximation: The successive convergents are

$$1, \frac{2}{3}, \frac{7}{10}, \frac{30}{43}, \frac{97}{139}$$

The convergents oscillate about the terminal value. The odd-numbered convergents (1, 7/10) are larger, the even numbered (2/3, 30/43) are smaller than the terminal fraction 97/139.

A simple method of calculating successive convergents makes use of the divisors of the original division sum. We illustrate with the example given above (p. 138).

To obtain the numerator of the fourth convergent (30), multiply the numerator of the third convergent (7) by the fourth divisor in the division sum (4) and add the numerator of the second convergent (2): $7 \times 4 + 2 = 30$. The denominator is obtained by a similar process.

An important property of convergents which is useful as a check on accuracy in their calculation may be expressed thus:

If p_{n-1}/q_{n-1} and p_n/q_n are successive convergents, then

$$p_n q_{n-1} - p_{n-1} q_n = (-1)^n$$

The reader may test this on the convergents of 97/139.

CONVERGENTS AND THE FIBONACCI SERIES

For comparison, the following continued fractions are of interest:

$$\sqrt{2} = 1 + \frac{1}{2+} \frac{1}{2+} \frac{1}{2+} \cdots$$

$$e = 2 + \frac{1}{1+} \frac{1}{2+} \frac{1}{1+} \frac{1}{1+} \frac{1}{4+} \frac{1}{1+} \frac{1}{1+} \frac{1}{6+} \cdots$$

$$\frac{\pi}{4} = \frac{1}{1+} \frac{1^2}{2+} \frac{3^2}{2+} \frac{5^2}{2+} \frac{7^2}{2+} \frac{9^2}{2+} \frac{11^2}{2+} \frac{13^2}{2+} \cdots$$

And now, consider the simplest of all infinitely continued fractions

$$1 + \cfrac{1}{1 + \cfrac{1}{1 + \cfrac{1}{1 + \cdots}}}$$

which is formally, but more conveniently written thus:

$$1 + \frac{1}{1+} \frac{1}{1+} \frac{1}{1+} \frac{1}{1+} \frac{1}{1+} \cdots$$

Forming the successive convergents, we obtain

$$1, \frac{2}{2}, \frac{3}{2}, \frac{5}{3}, \frac{8}{5}, \frac{13}{8}, \frac{21}{13}, \frac{34}{21}, \frac{55}{34}, \frac{89}{55}, \frac{144}{89}, \cdots$$

wherein both numerators and denominators constitute the Fibonacci series! The successive convergents oscillate about a value to which the series tends as a limit, that limit being the golden ratio, *Phi*.

$$\phi = 1 + \frac{1}{1+} \frac{1}{1+} \frac{1}{1+} \frac{1}{1+} \frac{1}{1+} \frac{1}{1+}$$

This remarkable result, beautiful in its neat simplicity, brings the golden section and the Fibonacci series into the closest possible association and stands as a worthy addition to our anthology.

The Fibonacci Numbers

We met the Fibonacci number series in chapter IV, p. 46. First discussed by Leonardo of Pisa (Fibonacci), they occupy today a place in mathematics which has justified the publication of a journal with the title *The Fibonacci Quarterly*. The first two members of the series, it will be recalled, are $u_1 = 1$ and $u_2 = 1$; thereafter each member is the sum of the two preceding members: $u_3 = u_1 + u_2 = 2$, $u_4 = u_2 + u_3 = 3$, and, in general, $u_{n+1} = u_{n-1} + u_n$. We shall see that this simple series has interesting properties.

In accordance with our declared intention of illustrating our main topic with exercises and examples which are confined to a narrow region of mathematics, in order to indicate with an anthology of very limited scope the depth of the soil from which a considerable mathematical harvest may be gathered, we continue to restrict ourselves to topics related to the golden section of the ancient Greeks.

We have seen that there is a connection between *Phi*, the golden ratio and the Fibonacci series. We have also found that an approximate value of *Phi* is obtained by dividing any Fibonacci number by the number of the series which precedes it:

$$\frac{u_{n+1}}{u_n} \approx \phi$$

In this chapter we shall show that any member of the Fibonacci series may be expressed accurately in terms of *Phi*:

$$u_n = \frac{\phi^n - \phi'^n}{\sqrt{5}}$$

The ubiquity of the golden ratio has been noted: it pops up unexpectedly in many different contexts. This is also characteristic of the Fibonacci numbers. Their appearance in the solution of combinatorial problems, as on p. 135, may not surprise us, but their emergence in a beehive or a rabbit warren, in a sunflower or a sea shell, is certainly unexpected. Such surprises cause us to inquire, introspectively, into the causes of the aesthetic feeling which they kindle. They are inevitably complex, but let us endeavor to disentangle two.

SOURCE OF AESTHETIC FEELING

First, there is the alternation of tension and relief, of perplexity and illumination. In the effort to resolve a run-of-the-mill mathematics problem, such as a text-book exercise, there is (as we have already noticed) a feeling of stress, of mental tension, almost of anxiety, so long as the perplexity continues. Then, with the solution of the problem, there comes relaxation. This relief is part of the pleasure of the discipline. It is not unknown, in a mathematics seminar, to hear the peace broken by a muttered expletive, "Got it!" which everyone present understands. Tension and relief: this is an emotional experience familiar in music as well as in mathematics. It ensues with the resolution of discord into harmony, or when the sound of the tonic replaces that of the dominant. This experience of illumination-succeeding-perplexity is one of the broad archaic experiences of mankind, its roots buried in the unconscious.

Second, the unexpected meeting with the Fibonacci numbers in an improbable context such as a beehive or in the expansion of an algebraic fraction, $x/(1 - x - x^2)$, is another aspect of the pleasure of the discipline of mathematics. More than a feeling of pleased surprise at the sudden encounter with a familiar friend, there is a sense of amazement: "How on earth did *you* get *here*?!"

The world of mathematics seems to have more interconnecting paths than we had thought. Moreover, discerning minds experience a sense of wonder: how remarkable that so simple a series, derived from so simple a rule, should have so many affiliations!

Such emotions, archaic in origin, contribute to the charm of mathematics. The experience of glimpsing beauty in mathematics is as difficult to interpret to oneself as it is to communicate to a pupil. It is caught rather than taught. The student can only be encouraged to see the "vision splendid" for himself. The joy, mediated through the intellect, originates in lower strata of the mind, the arena of the emotions. Poincaré wrote:

It may appear surprising that sensibility should be introduced in connection with mathematical demonstrations, which, it would seem, can only interest the intellect. But not if we bear in mind the feeling of mathematical beauty, of the harmony of numbers and forms and of geometric elegance. It is a real aesthetic feeling that all true mathematicians recognize, and this is truly sensibility.... The useful combinations are precisely the most beautiful, I mean those that can most charm that special sensibility that all mathematicians know, but of which laymen are so ignorant that they are often tempted to smile at it.[1]

◘   ◘   ◘

There follows a list of the first forty members of the Fibonacci sequence:

THE FIBONACCI NUMBERS

u_1	1	u_{11}	89	u_{21}	10946	u_{31}	1346269
u_2	1	u_{12}	144	u_{22}	17711	u_{32}	2178309
u_3	2	u_{13}	233	u_{23}	28657	u_{33}	3524578
u_4	3	u_{14}	377	u_{24}	46368	u_{34}	5702887
u_5	5	u_{15}	610	u_{25}	75025	u_{35}	9227465
u_6	8	u_{16}	987	u_{26}	121393	u_{36}	14930352
u_7	13	u_{17}	1597	u_{27}	196418	u_{37}	24157817
u_8	21	u_{18}	2584	u_{28}	317811	u_{38}	39088169
u_9	34	u_{19}	4181	u_{29}	514229	u_{39}	63245986
u_{10}	55	u_{20}	6765	u_{30}	832040	u_{40}	102334155

There are features of this table, not immediately apparent, which are worthy of note.

It may be shown, by continuing the table to higher values, that the figures in the units places form a recurring series with a period of 60:

u_0	0	u_{60}	$\cdots 20$
u_1	1	u_{61}	$\cdots 61$
u_2	1	u_{62}	$\cdots 81$
u_3	2	u_{63}	$\cdots 42$
u_4	3	u_{64}	$\cdots 23$
u_5	5	u_{65}	$\cdots 65$
$\cdots$	$\cdots$	$\cdots$	$\cdots\cdots$

The figures in the tens places have recently been shown also to constitute a recurring series but with a period longer than 60. Still more recently, with the aid of a computer, the figures in the hundreds and in the thousands places have been shown to recur, but with periods of impracticable length.

It is interesting that, if the series is expressed in the binary scale, the period is 24 in the first four places, i.e., those denoting $2^0, 2^1, 2^2, 2^3$.

We have seen that the golden ratio $\phi \to u_{n+1}/u_n$ as $n \to \infty$. Taking n as small as 30, using the above table and a desk calculator, we obtain

$$u_{30}/u_{29} = 1.618033988749874831 \cdots$$

Even with seven-place logarithms we obtain an approximation sufficient for most purposes:

DIFFERENCES

$$\log u_{31} = \log 1346269 = 6.1291322$$
$$\log u_{30} = \log 832040 = 5.9201442 \qquad 0.2089880$$
$$\log u_{29} = \log 514229 = 5.711567 \qquad 0.2087857$$

FIBONACCI SERIES: A GEOMETRIC PROGRESSION

From these figures we not only obtain a value of *Phi* to four decimal places, but we discover a surprising fact about the Fibonacci series. True, it is an empirical result, but we shall shortly be able to confirm it rigorously on theoretical grounds.

From the above figures we obtain

$$\frac{u_{31}}{u_{30}} \approx 1.61803(5) \approx \frac{u_{30}}{u_{29}}$$

Thus, the interesting result emerges that, as the terms of the series increase, the Fibonacci sequence approximates ever more closely to a geometrical progression, the common ratio of the series being none other than the golden ratio, *Phi*. These pleasing

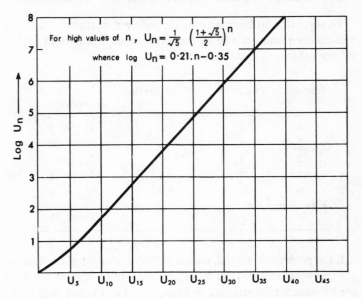

Fig. 11.1. Graph of log u_n

results will be confirmed when we meet with Binet's formula for the general term of the Fibonacci series. We shall find that, for large values of n,

$$u_n \approx \frac{1}{\sqrt{5}} \left(\frac{1 + \sqrt{5}}{2}\right)^n$$

whence

$$\log u_n \approx 0.21n - 0.35$$

Figure 11.1 shows a graphical expression of the equation.

Before we investigate other sources of the Fibonacci numbers, let us consider a practical problem which serves to show that these numbers turn up in unlikely places.

> PROBLEM: *An endowment produced $9 per week for an orphanage school to pay for seats at a local concert hall. The tickets cost $2 each for a teacher and $1 each for a child. In how many ways could a party for the hall be arranged, assuming that neither an interchange of teachers nor an interchange of children, but only an interchange of teacher and child produced a new arrangement.*

To solve this problem we note the different mixtures of teachers and children which can be allowed within the price limit of $9. We tabulate this as follows:

TEACHERS	CHILDREN	NO. OF SEATS	NO. OF ARRANGEMENTS
4	1	5	$^{5}C_1 = 5$
3	3	6	$^{6}C_3 = 20$
2	5	7	$^{7}C_5 = 21$
1	7	8	$^{8}C_7 = 8$
0	9	9	$^{9}C_9 = 1$
			Total 55

If the problem is altered, making $12 the price limit, the number of arrangements turns out to be 233. Now both of these numbers are Fibonacci numbers: $u_{10} = 55$, $u_{13} = 233$. It is not difficult to show that, whatever the price limit may be, the answer will be a Fibonacci number. Suppose the price limit is $2n$ dollars. The tabulation is now as follows:

TEACHERS	CHILDREN	NO. OF SEATS	NO. OF ARRANGEMENTS
n	0	n	1
$n - 1$	2	$n + 1$	$^{n+1}C_2$
$n - 2$	4	$n + 2$	$^{n+2}C_4$
...	...	...	...
0	$2n$	$2n$	$^{2n}C_{2n}$

The total is:

$$1 + {}^{n+1}C_2 + {}^{n+2}C_4 + \cdots + {}^{2n-1}C_{2n-2}$$

Reference to the broken line of the Chinese triangle (p. 133) will show that this summation leads to a Fibonacci number, e.g., the triangle shows that

$${}^5C_1 + {}^6C_3 + {}^7C_5 + {}^5C_7 + {}^9C_9 = 55$$

ANOTHER SOURCE

The Fibonacci numbers appear again in a quite different context. Let us find the first few coefficients of the expansion of $x/(1 - x - x^2)$ by direct division. The quotient is

$$x + x^2 + 2x^3 + 5x^5 + 8x^6 + 13x^7 + 21x^8 + \cdots$$

These coefficients form the Fibonacci series.

The first few coefficients of the expansion of $x/(1 + x - x^2)$ are the same except that they are alternately positive and negative:

$$1, -1, +2, -3, +5, -8, +13, -21, \cdots$$

Let us find the value of the coefficient of x^n in the expansion of $x/(1 - x - x^2)$, which is u_n.

Write $x/(1 - x - x^2) = A/(x - \alpha) + B/(x - \beta)$, where α, β are the roots of $1 - x - x^2 = 0$, so that $\alpha\beta = 1$ and $\alpha + \beta = 1$.

From this we have $x = A(x - \alpha) + B(x - \beta)$, whence $A = \alpha/(\alpha - \beta)$, $B = \beta/(\beta - \alpha)$. Thus, we require the coefficients of x^n in the expansion of

$$\frac{1}{\alpha - \beta}\left(\frac{\alpha}{x - \alpha} - \frac{\beta}{x - \beta}\right) \qquad \text{(i)}$$

Consider the first term $\alpha/(x - \alpha) = -(1 - x/\alpha)^{-1} = -(1 - t)^{-1}$, where $t = x/\alpha$. But

$$-(1 - t)^{-1} = -(1 + t + t^2 + t^3 + \cdots + \cdots)$$

The term containing x^n is $t^n = -x^n/\alpha^n$. Hence, the coefficient of x^n in the expansion is, from (i), $1/(\alpha - \beta)(-1/\alpha^n + 1/\beta^n)$, i.e.,

$$\frac{\alpha^n - \beta^n}{\alpha - \beta} \cdot \frac{1}{\alpha^n \beta^n} = (-1)^n \frac{\alpha^n - \beta^n}{\alpha - \beta} = u_n$$

Now the roots of $x^2 - x - 1 = 0$ are

$$\alpha = \frac{1 + \sqrt{5}}{2}, \quad \beta = \frac{1 - \sqrt{5}}{2} \qquad \text{(ii)}$$

BINET'S FORMULA

We are now in a position to find the value of the continued fraction

$$1 + \frac{1}{1+} \frac{1}{1+} \frac{1}{1+} \frac{1}{1+} \cdots$$

It is the convergent u_{n+1}/u_n as $n \to \infty$. Now

$$\frac{u_{n+1}}{u_n} = (-1)^{n+1} \cdot \frac{\alpha^{n+1} - \beta^{n+1}}{\alpha - \beta} \bigg/ (-1)^n \frac{\alpha^n - \beta^n}{\alpha - \beta}$$

i.e.,

$$\frac{u_{n+1}}{u_n} = -\frac{\alpha^{n+1} - \beta^{n+1}}{\alpha^n - \beta^n}$$

Since $\beta < 1$, $\lim_{n \to \infty} \beta^n = 0$; hence $\lim_{n \to \infty} u_{n+1}/u_n = \alpha$. This limit we have called ϕ, so that $\phi \equiv \alpha = (1 + \sqrt{5})/2$. (The other expansion $x/(1 + x - x^2)$ gives the same value for ϕ with the opposite sign.)

From (ii), since $\alpha - \beta = \sqrt{5}$, we may write the $(n + 1)$th term of the Fibonacci series:

$$u_n = \frac{1}{\sqrt{5}} \left[\left(\frac{1 + \sqrt{5}}{2} \right)^n - \left(\frac{1 - \sqrt{5}}{2} \right)^n \right] \quad (n = 0, 1, 2, \cdots)$$

This is Binet's formula. If n is large, the second term is negligible, and

$$u_n \approx \frac{1}{\sqrt{5}} \left(1 + \frac{\sqrt{5}}{2} \right)^n$$

Writing $\beta = (1 - \sqrt{5})/2 = \phi'$,

$$u_n = \frac{1}{\sqrt{5}} \left(\phi^n - \phi'^n \right)$$

SOME PROPERTIES OF FIBONACCI NUMBERS

The Fibonacci numbers exhibit a variety of curious properties and this chapter would not be complete without a reference to some of them. A brief anthology has space for a few only, but the reader may take it that, for every one recorded here, twenty may be found in the literature of the subject. This is an indication of the fertility of the soil in this area of mathematics. But this is not our main objective, which is rather to provide further specimens of beauty in mathematics. The following theorems have qualities that can evoke aesthetic appreciation in a mind trained to assess mathematical values—neatness, brevity, elegance and an unexpectedness which may stimulate surprise, and even a sense of wonder.

1. We have

$$u_1 = u_3 - u_2 = u_3 - 1$$
$$u_2 = u_4 - u_3$$
$$\cdot \quad \cdot \quad \cdot \quad \cdot \quad \cdot \quad \cdot \quad \cdot$$
$$u_{n-1} = u_{n+1} - u_n$$
$$u_n = u_{n+2} - u_{n+1}$$

By addition, $u_1 + u_2 + \cdots + u_n = u_{n+2} - 1$.

2. The following results may be proved by a similar method:

i. $u_2 + u_4 + \cdots + u_{2n} = u_{2n+1} - 1$

ii. $u_1 + u_3 + \cdots + u_{2n-1} = u_{2n}$

3. Since $u_r u_{r+1} - u_{r-1} u_r = u_r(u_{r+1} - u_{r-1}) = u_r^2$, we have

$$u_1^2 = u_1 u_2$$
$$u_2^2 = u_2 u_3 - u_1 u_2$$
$$u_3^2 = u_3 u_4 - u_2 u_3$$
$$\cdot \quad \cdot \quad \cdot \quad \cdot \quad \cdot \quad \cdot \quad \cdot$$
$$u_n^2 = u_n u_{n+1} - u_{n-1} u_n$$

By addition, $u_1^2 + u_2^2 + \cdots + u_n^2 = u_n u_{n+1}$.

4. The following results may be established by the method of mathematical induction:

i. The square of a Fibonacci number and the product of the

number that precedes it and the number that follows it in the series differ by unity: $u_{n+1}^2 = u_n u_{n+2} + (-1)^n$.

ii. The difference of the squares of two Fibonacci numbers whose subscripts differ by two is a Fibonacci number:

$$u_{n+1}^2 - u_{n-1}^2 = u_{2n}$$

5. If r is any integer, u_n is a factor of u_{rn}.

Example. $u_{11} = 89$ is a factor of $u_{22} = 17711 = 89 \times 199$ and of $u_{33} = 3524578 = 89 \times 3962$ (see p. 143 for a list of Fibonacci numbers).

6. If d is the highest common factor of m and n, then u_d is the highest common factor of u_m and u_n.

Example. The highest common factor of $u_{14} = 377$ and $u_{21} = 10946$ is $u_7 = 13$, for $377 = 13 \times 29$ and $10946 = 13 \times 942$.

7. If m and n have no common factor, u_{mn} is divisible by $u_m \times u_n$ without remainder.

Example. $u_{28} = 317811 = 39 \times 8149 = u_4 \times u_7 \times 8149$.

FIBONACCI NUMBERS YIELD ZERO-VALUE DETERMINANTS

As a tail-piece to this chapter it may be remarked that a determinant formed of successive Fibonacci numbers has the value zero. For example:

$$\begin{vmatrix} 3 & 5 & 8 \\ 13 & 21 & 34 \\ 55 & 89 & 144 \end{vmatrix} = 0 \qquad \begin{vmatrix} 1 & 2 & 3 & 5 \\ 8 & 13 & 21 & 34 \\ 55 & 89 & 144 & 233 \\ 377 & 610 & 987 & 1597 \end{vmatrix} = 0$$

This follows from the law of the sequence: $u_{n-1} + u_n = u_{n+1}$, for the addition of adjacent columns produces the next column, making the determinant zero.

CHAPTER XII

Nature's Golden Numbers

In former pages we have hinted that Nature herself is familiar with the golden section and its near relative, the Fibonacci sequence. The famous rabbit breeding problem of ancient days (p. 158) is an example. In this chapter we consider further examples, such as the genealogy of the drone bee, conifers, the shells of mollusks, flowers of the composite family such as the sunflower, and phyllotaxis. Thus the mathematical framework of beauty found in Nature may be invoked to supplement our anthology.

It is important to bear in mind that Nature's surface beauty conveys no more than a hint of the loveliness hidden within. The mathematics is not in its skin: it must be uncovered. This spells labor. The poets for the most part fail to reckon with this. Indeed, their view-point is sometimes so mistaken that it is worth while to spend a moment in refuting it, the more so as it is a principal aim of this essay to disseminate the opposite view.

Few, if any, of the English major poets were moved more deeply than Wordsworth by the sights and sounds of Nature; no one had greater skill in communicating his emotion by the written word. His verse is at times almost unbearably moving. But consider the limitations.

He began one of his poems thus:

> My heart leaps up when I behold
> A rainbow in the sky . . .

So does mine. Not, I am sure, with so intense a feeling as did Wordsworth's. But it takes a longer leap! You see, one of the loveliest topics in the whole range of the discipline of physics is concerned with the hidden mechanism of the rainbow. It is a feast of delight! For the scientifically literate, an inspired writer might produce a large book on the rainbow to which the reader would react as one does to poetry. For the scientifically illiterate, however, the simple rules of the reflection of light and of diffraction, Snell's laws applied to the falling raindrops, the beautiful geometry of their spherical form and other such delightful, cognate topics relating to the rainbow make no appeal at all. For the mathematician of even modest attainments there is here a rich harvest of mathematical beauty, the contemplation of which can induce the feeling—indeed, the conviction—in those who are educated to appreciate it, that they are "thinking God's thoughts after Him."

But Wordsworth would have none of this! In his poem "The Tables Turned" which begins "Up! Up! my Friend, and quit your books," he concludes:

> Sweet is the lore which Nature brings;
> Our meddling intellect
> Mis-shapes the beauteous forms of things:—
> We murder to dissect.
>
> Enough of Science and of Art;
> Close up those barren leaves;
> Come forth, and bring with you a heart
> That watches and receives.

The poet Blake expresses somewhat similar sentiments when (in a letter to Butts, 22 Nov. 1802) he prays to be delivered from "single vision and Newton's sleep."

Newton was not asleep, but wide awake to some of the loveliest examples of natural beauty known to man. When we note the perfection of the circular form of (say) a lunar halo, or the shape of the parabola outlined by a jet of water, or that of the cycloid described by a point on a wheel or many such naturally occurring curves, we experience more than the immediate satisfaction which Blake himself must have had in such curves; we derive additional pleasure from the beauty which Newton's genius uncovered in these familiar shapes.

TELEOLOGY

Reverting for a moment to the rainbow, it may seem naïve to ask, "What is a rainbow for?" as though every creation required a teleological interpretation. But if its origin was the deliberate *fiat* of a creator, since its end is certainly not utilitarian, one can hardly help wondering whether it was intended solely to make the heart leap up in Wordsworth and in others like him.

I can find no answer to this question, and J. B. Mozley's comment reveals a comparable inadequacy:

The beauty of nature is a distinct revelation made in the human mind, apart from that of its use. When the materialist has exhausted himself to explain utility in nature, it would appear to be the peculiar office of beauty to rise up suddenly as a confounding and baffling extra, which was not even formally provided for in his scheme. The glory of nature resides in the mind of man. We cannot explain why material objects impress the imagination. The whole of what any scene of earth or sky is *materially* is stamped upon the retina of the brute, just as it is upon man's; the brute sees all the same objects which are beautiful to man— only *without their beauty*; which aspect is inherent in man, and part of his reason.

FIBONACCI AND NATURE

The problem of the function of beauty in the human milieu is difficult; but, if one asks what it achieves, a part of the answer is that it serves as a lure to induce the mind to embark on creative activity. Beauty is a bait. This view seems to require the existence of "absolute" beauty, to demand that specimens of beauty antedate the human perception of them, although beauty in its subjective sense is called into existence only at the moment of its appreciation. This contradicts a widely held opinion, held for example by Morris Kline, who writes concerning beauty in mathematics:

> It would appear as though mathematics is the creation of human fallible minds rather than a fixed, externally existing body of knowledge. The subject seems very much dependent on the creator.[1]

Kline quotes A. N. Whitehead in support of this point of view: "The science of pure mathematics may claim to be the most original creation of the human spirit."

This view is difficult to reconcile with that of Sir James Jeans who appeared to affirm that mathematics is indeed an "eternally existing body of knowledge," when he uttered his well-remembered aphorism that "God is a mathematician."

In this chapter examples of *sectio divina* and Fibonacci sequences found in Nature will provide opportunities to examine instances that throw light on this discussion. We shall find a simple example in the fact that the great-great-grandparents of a drone bee number five bees—no more, no less—and the parents of these ancestors number eight. Now 5, 8, and $5 + 8 = 13$, which is the number of all the ancestors' grandparents, are all numbers of the Fibonacci series, which (as we have seen) has interesting mathematical properties. These were facts a million years before Fibonacci was born. No mathematician created them. Someone discovered them and expressed them in mathematical symbols, again "thinking God's thoughts after Him."

Before we consider examples from the biological sciences let us look at a couple drawn from the world of inorganic phenomena.

MULTIPLE REFLECTIONS

If a beam of light is incident upon two sheets of glass in contact, part of the light will be transmitted, part absorbed and the remainder reflected. There will be multiple reflections. The number of different paths followed within the glass before the ray emerges depends on the number of reflections which the ray undergoes. The number of emergent rays is a *Fibonacci number*. This is indicated in figure 12.1:

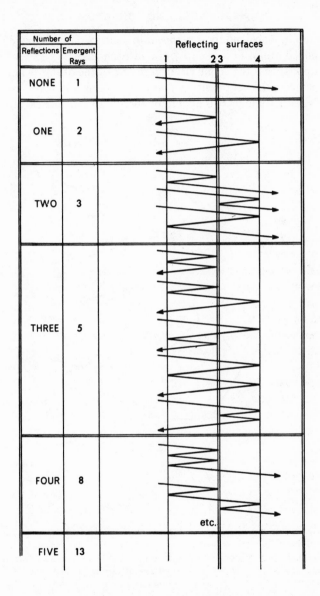

Fig. 12.1. Multiple reflections

FIBONACCI AND THE ATOM

The Fibonacci numbers reappear in connection with the ideally simplified atoms of a quantity of hydrogen gas.

Suppose that the single electron in one of the atoms is initially in the ground level of energy and that it gains and loses, successively, either one or two quanta of energy, so that the electron in

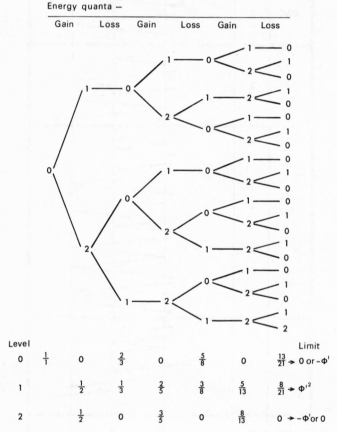

Fig. 12.2. Possible histories of an atomic electron. These fractions, formed of Fibonacci numbers, show the proportion of atoms in each state as time passes. The number of possible different histories of an electron are also members of the Fibonacci sequence:

$$1 \quad 2 \quad 3 \quad 5 \quad 8 \quad 13 \quad 21 \quad \cdots$$

its history occupies either the ground level (state 0), the first energy level (state 1), or the second energy level (state 2). In this idealized case the number of different possible histories of an atomic electron is a Fibonacci number (Fig. 12.2, p. 156).

Let us make the following assumptions:

1. When the gas gains radiant energy, all the atoms in state 1 rise to state 2; half the atoms in state zero rise to state 1 and half to state 2.

2. When the gas loses energy by radiation, all the atoms in state 1 fall to state zero; half those in state 2 fall to state 1 and half to state zero.

Figure 12.2 shows the successive fractions of the total number of atoms found in each state. These fractions are formed exclusively of Fibonacci numbers.

A point of interest is that the fraction of atoms in the intermediate energy level (state 1) remains constant at 38.2 per cent. If u_n is the nth term of the Fibonacci series, this fraction (38.2 per cent) is u_n/u_{n+2} as n tends to infinity.

$$\frac{u_n}{u_{n+2}} = 1 - \frac{u_{n+1}}{u_{n+2}} \rightarrow 1 + \phi'$$

i.e., 38.2 per cent. The symbols ϕ and ϕ' stand for the limits of u_{n+1}/u_n and u_n/u_{n+1} respectively as n tends to infinity. We have seen that they are the roots of the equation $x^2 - x - 1 = 0$.

◘ ◘ ◘

LEONARDO OF PISA

Before we describe how the golden numbers emerge in biological references by discussing the rabbit breeding problem, a note about the originator of this ancient and famous problem will be of interest.

Leonardo Fibonacci (filius Bonacci), alias Leonardo of Pisa, was born nearly 800 years ago—in A.D. 1175. His early years were passed in a Christian community, but he received his academic education among the Mohammedans of Barbary. There he learned the Arabic, or decimal, system of numbering as well as Alkarismi's teaching of algebra. When about twenty-seven years of age he returned to his native land and there published a work which became widely known as *Liber Abaci* (the book of the Abacus), in which he demonstrated the great advantages of the Arabic system of numeration over the Roman. Today we need no convincing that it is easier to write 98 than XCVIII. *Liber Abaci* (1202), Fibonacci's *magnum opus* was a standard work for two hundred years and the principal means of introducing the Hindu-Arabic system of notation to the educated classes of Christian Europe.

Leonardo's reputation among scholars was deservedly great. It was so outstanding that Frederick II, visiting Pisa in 1225, held a public competition in mathematics there to test Leonardo's skill. A specimen problem was: "What number, when squared and either increased or decreased by 5, would still be a perfect square?" Leonardo gave the answer 41/12, which is correct, for

$$(41/12)^2 + 5 = (49/12)^2 \quad \text{and} \quad (41/12)^2 - 5 = (31/12)^2$$

Leonardo's competitors did not succeed in solving any of the problems set.

THE RABBIT PROBLEM

As we have remarked, we find the Fibonacci series (like the golden section) cropping up in unexpected places. Who would have expected it to have a connection with the breeding of rabbits?! Long before myxamatosis solved the rabbit problem which was one of the agricultural headaches of the Australian farmer, mathematicians solved the rabbit problem which had bothered the mathematical contemporaries of Fibonacci.

The Fibonacci series originated in a mathematical puzzle proposed by Fibonacci in *Liber Abaci*. He proposed that the progeny of a single pair of rabbits arrived as follows.

Suppose there is one pair of rabbits in the months of January which breed a second pair in the month of February and that thereafter these produce another pair monthly, that each pair of rabbits produce another pair in the second month following birth and thereafter one pair per month.

The problem is to find the number of pairs at the end of the following December.

To solve this puzzle we tabulate in four columns:

1. Number of pairs of breeding rabbits at the beginning of given month.

2. Number of pairs of non-breeding rabbits at the beginning of the month.

3. Number of pairs of rabbits bred during the month.

4. Number of pairs of rabbits living at the end of the month.

MONTH	1	2	3	4
January	0	1	0	1
February	1	0	1	2
March	1	1	1	3
April	2	1	2	5
May	3	2	3	8
June	5	3	5	13
July	8	5	8	21
August	13	8	13	34
September	21	13	21	55
October	34	21	34	89
November	55	34	55	144
December	89	55	89	233

Each of these columns contains the Fibonacci series, formed according to the rule that any term is the sum of the two immediately preceding terms,

$$u_{n+1} = u_n + u_{n-1} \qquad (u_0 = 0, u_1 = 1)$$

THE BEE HIVE

In discussing in chapter IX patterns and designs found in Nature, we noted that one of the attractive patterns is found in the honeycomb. The cells of wax designed as honey receptacles are of hexagonal cross-section forming a continuous pattern which fills space without interstices. The only other simple way of achieving this is by cells of rectangular cross section, preferably square for the sake of rigidity.

Why do the bees choose the hexagonal pattern? If this is a

question in psychology, the answer is not forthcoming. But if it is a question in mathematics, the answer is that the shape is determined by consideration of economy and efficiency.

The honeycomb is a pattern in space. The genealogical table of a bee is a pattern in time. The Mathematician who participated in the former had a hand also in the latter. The drone, or male bee, hatches from an egg that has not been fertilized. The fertilized egg produces only females—queens or workers. If we use this fact of

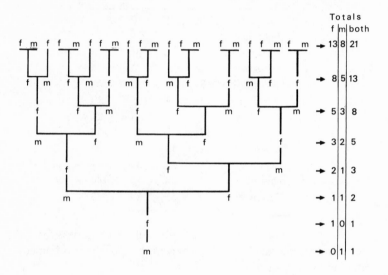

Fig. 12.3. Geneaology of drone bee

life to draw up a genealogical table showing the ancestry of a drone bee for several generations, we shall arrive at a diagram like that of figure 12.3. Making totals of all the males, all the females, and all the bees of both sexes that constitute each generation, we find that we have overlapping Fibonacci series thrice repeated—one for males, one for females, and one for both combined. This pretty result is displayed on the right-hand side of figure 12.3.

It is not only the zoologist through his rabbits and the entomologist through his bees who make contact with the golden numbers. The botanist also meets them in different areas of his studies—in leaf arrangement, in petal structure, in florets of the composite family, and in the arrangement of the axils on the stems of a plant. It is rather rare to find perfect specimens which conform accurately to the mathematical pattern. A field daisy may have 33 or perhaps 56 petals which just miss the Fibonacci numbers 34 and 55, but a daisy with a petal count between (say) 40 and 50 would be uncommon.

PHYLLOTAXIS

Phyllotaxis is the botanical term for a topic which includes the arrangement of leaves on the stems of plants. The arrangements are characteristic of the genera. Leaf "divergence" is the technical term used to describe the angular separation of two successive leaf bases on the stem as measured by a helix drawn from the root of the plant upwards to its growing point (Fig. 12.4). The leaf arrangement can be specified in terms of this divergence.

A helix is drawn to pass through each leaf base until it reaches the first base which is vertically above the starting point. Let p be the number of turns of the helix and q the number of leaf bases

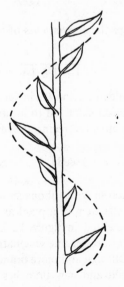

Fig. 12.4. Phyllotaxis

passed (excluding the first). Then p/q is a fraction which is characteristic of the plant, the leaf divergence. Both numerator and denominator of this fraction tend to be members of the Fibonacci sequence:

$$\frac{1}{2}, \frac{1}{3}, \frac{2}{5}, \frac{3}{8}, \frac{5}{13}, \frac{8}{21}, \cdots$$

The botanist's interest in leaf divergence is not primarily mathematical; his attention is directed rather to the fact that all the members of this series of fractions lie between 1/2 and 1/3, so that the successive leaves are separated from one another by at least one-third of the stem circumference, ensuring maximum illumination and air for each leaf base.

H. E. Licks states that, as a general rule, divergences for various plants can be arranged as follows:[2]

Common grasses	1/2
Sedges	1/3
Fruit trees, such as apple	2/5
Plantains	3/8
Leeks	5/13

These fractions are the convergents of the continuous fraction

$$\frac{1}{2} + \frac{1}{1+} \frac{1}{1+} \frac{1}{1+} \cdots$$

So we meet our golden section *Phi* in yet another connection, for this continuous fraction extended to an infinite number of terms converges to ϕ'^2 which is the reciprocal of ϕ^2. We have seen that

$$\phi = -\frac{1}{\phi'} = \frac{\phi + 1}{\phi} = 1.618034\cdots, \quad \text{and} \quad \phi'^2 = 0.38197\cdots$$

A different connection with Fibonacci numbers is found in the number of axils on the stem of a plant as it develops. An ideally simple case is represented in figure 12.5, where the stems and flowers of sneezewort are set out schematically. A new branch is seen to spring from the axil and more branches grow from the new branch. Since the old and new branches are added together, a Fibonacci number is found in *each horizontal plane*.

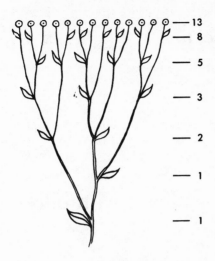

Fig. 12.5. Sneezewort (*Achillea ptarmica*)

The golden numbers come into view again if we examine the number of petals of certain common flowers; examples are:

Iris	3 petals
Primrose	5 petals
Ragwort	13 petals
Daisy	34 petals
Michaelmas daisy	55 and 89 petals

The number and arrangement of the florets in the head of a member of the composite family is a particularly beautiful example of golden numbers found in Nature. We reserve this for the next chapter.

The emergence in the natural world of the terms of the Fibonacci sequence is easily understood in some of the examples we have described, as, for instance, the genealogy of the drone bee. In other cases, however, like those known to the botanist, the biological explanation is not so easy to see. It is one of the surprises of mathematics that its results, often reached in complete isolation from the phenomenal world, can prove to be closely applicable to it.

CHAPTER XIII

Spira Mirabilis

To conclude this brief anthology of the golden section, we will consider in greater detail the beautiful curve, already mentioned in chapter VII, which has not only been studied by mathematicians for hundreds of years but has been represented in Nature for thousands of centuries in flora and in fauna: the equiangular spiral. Many people will have seen this curve, though most of them will not have noticed it, in the pattern of the florets of the common daisy. It is more easily observed in another member of the composite family—the sunflower. In a good specimen a remarkable feature will be seen: two sets of equiangular spirals superposed or intertwined, one being a right-handed, and the other a left-handed spiral, each floret filling a dual role by belonging to both spirals. This fascinating combination, represented in figure 13.1, is remarkable enough, but it is astonishing to learn that the numbers of spirals are adjacent Fibonacci numbers: there are 21 clockwise, and 34 anti-clockwise spirals! This combination of (i) the charm of the opposing spiral patterns, (ii) the dual role of each floret, (iii) the simplicity of the mathematical representation, and (iv) the unexpected association with the familiar Fibonacci series (not to mention color outline),

Fig. 13.1. Spiral pattern in the sunflower

together constitute a most impressive example of mathematical beauty drawn from Nature.

Comparable arrangements of opposing spirals associated with Fibonacci numbers are found in the pine cone (5 and 8) and in the pineapple (8 and 13).

SHELLS

Outstanding examples of the equiangular spirals found in Nature are the shells of a wide variety of creatures, from the diminutive foraminifera, which would pass through the eye of a needle, to the nautilus, measuring several centimeters in diameter. The beauty of the chambered nautilus has attracted the attention and the admiration of the mathematician, the zoologist, the palaeontologist, the artist, and the poet.

The marine zoologist says that

...the pearly Nautilus has an internal spiral with dozens of little chambers divided off by walls of purest mother of pearl inside the shell. As the animal grows and extends the mouth of the spiral shell, it every now and again moves forward into more commodious quarters, closing the door behind it with a layer of nacre, or pearl, so that it occupies only the outermost chamber, and, naturally, each successive chamber is larger than the last. The chambered part of the shell is filled with gas or air, so that the whole thing remains buoyant in spite of its massive build. A little thin tail, or siphuncle, extends from the hind end of the animal right through to the baby beginning of the shell, passing through smooth holes left in the partitions....[1]

Oliver Wendell Holmes was inspired by the nautilus to write a poem which he entitled "The Chambered Nautilus":

> This is the ship of pearl, which, poets feign,
>   Sails the unshadowed main—
> The venturous bark that flings
>   On the sweet summer wind its purpled wings
> In gulfs enchanted, where the siren sings,
>   And coral reefs lie bare,
> Where the cold sea-maids rise to sun their streaming hair.

Although, according to Crosbie Morrison, the poet has "got his mollusks mixed," he knows enough of their life history to draw a moral in his last stanza:

> Build thee more stately mansions, O my soul,
>   As the swift seasons roll!
> Leave thy low-vaulted past!
>   Let each new temple, nobler than the last,
> Shut thee from heaven with a dome more vast
>   Till thou at last art free,
> Leaving thine outgrown shell by life's unresting sea.

EXPRESSIVE LINES

The pearly nautilus attracts the artist both by the tints of its lustrous exterior and by the perfection of its spiral curve. The latter feature must be counted as an ingredient of the mathematical beauty of the equiangular spiral. We are reminded once again that aesthetic appreciation of any sort has a dual aspect. The immediate sensuous pleasure evoked by beauty is a common human experience: it is inborn, a natural endowment; but this rudimentary satisfaction can be developed by education.

The sensuous satisfaction (and its opposite) which is produced by simple lines has been studied by psychologists. The following is an account of some of Lundholm's experiments in this area:

Expressiveness of lines. When asked to draw a beautiful line, Lundholm's (1921) subjects tried to make one that was smooth, curved, symmetrical continuous, with rhythm or repetition and expressive of a single idea. For an ugly line they drew an unorganized mass without continuity, with mixed angles and curves and unrelated spaces...and, when they wished to express merriment, playfulness, agitation or fury, they drew sharp waves or zigzags.

The subjects said: "Small waves make the movement of a line go more quickly. The calm line has long slow curves."

Another: "Angularity of a line expresses violence of movement.... A long slow curve always expresses slowness."[2]

"The calm line has long slow curves." The long slow curve of the equiangular spiral, according to this, must be evocative of calm feelings which may be regarded as a part of the mathematician's aesthetic experience.

AN X-RAY PICTURE

The human race has never in its history been without opportunities to observe the "long slow curve" of the mollusk. Nautiloid mollusks were plentiful 400 million years ago. Some of these closely resembled their surviving relative the nautilus, which is still plentiful in the ocean north of the Fiji Islands.

Fossils of foraminifera, prolific unicellular marine organisms, with diameters of the order of a millimeter, are used to date rock strata by geologists in search of oil. A Kodak research team has shown that it is possible, by using an x-ray technique and special fine-grained x-ray film, to photograph the internal structure of these minute fossils without dissecting or cutting them in any way. These reveal that the identical equiangular spiral structure has persisted for scores of millions of years, with small variations that make dating possible. An x-ray photograph of a chambered *Nautilus pompilius*, about six inches long, is shown on the Frontispiece.

VARIOUS DESIGNATIONS

The spiral has been known by a variety of names corresponding to one or another of its features. By Descartes, who discussed it in 1638, it was designated the *equiangular spiral*, because the angle at which a radius vector cuts the curve at any point is constant. Because its radius increases in geometrical progression as its polar angle increases in arithmetical progression, it has been called the *geometrical spiral*. Halley, noting that the lengths of the segments cut off from a fixed radius by successive turns of the curve were in continued proportion, named it the *proportional spiral*. Jakob Bernoulli (1654–1705), who was so fascinated by the mathematical beauty of the curve that he asked that it might be engraved on his tombstone, called it (for reasons that will be stated later) the *logarithmic spiral*.

In more recent times, Rev. H. Moseley, a Canon of Bristol Cathedral (grandfather of the brilliant young physicist of Atomic Number fame, killed in Gallipoli in 1915) gave a simple, mathematical account of the spiral shell. Even earlier, Sir Christopher Wren, considering its architecture, perceived that the spiral was a cone coiled about an axis.

The fundamental mathematical property of the equiangular (or logarithmic) spiral corresponds precisely to the biological principle that governs the growth of the mollusk's shell. This principle is the simplest possible: the size increases but the shape is unaltered. The mollusk's shell grows longer and wider to accommodate the growing animal, but the shell remains always similar to itself. It grows at one end only, each increment of length being balanced by a proportional increase of radius so that its form is unchanged. The shell grows by accretion of material; more accurately, it accumulates rather than grows.

The only mathematical curve to follow this pattern of growth is the logarithmic spiral. Because of this, Bernoulli described it as *spira mirabilis*. Of course, the pattern of development can be imitated by mathematical forms other than this spiral. A rectangle, a parallelogram, a cone, etc. can grow while remaining similar to itself in shape. This interested not only the Greeks of 500 B.C. but also the Egyptians a thousand years earlier.

GNOMONS

This suggests a more general approach to the logarithmic spiral, via the ancient concept of the *gnomon*, an example of which was seen in chapter VII (Fig. 7.6). A gnomon is a portion of a figure which has been added to another figure so that the whole

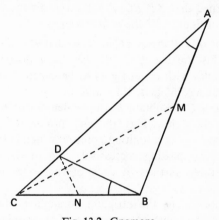

Fig. 13.2. Gnomons

is of the same shape as the smaller figure. Hero of Alexandria showed that in any triangle *ABC* (Fig. 13.2), triangle *ABD* is a gnomon to triangle *BCD* if $\angle CBD = \angle A$. If we add to, or subtract from this triangle a series of gnomons, it turns out that all the apices lie upon an equiangular spiral.

Radial growth (*dr*) and intrinsic growth in the direction of the curve (*ds*) bear a constant ratio to each other: $dr/ds = \cos \alpha =$ constant. We have seen that the equiangular spiral is the only curve to possess this property.

D'Arcy Thompson writes:

In the growth of a shell we can conceive no simpler law than this, namely, that it shall widen and lengthen in the same unvarying proportions: and this simplest of laws is that which Nature tends to follow.

The shell, like the creature within it, grows in size *but does not change its shape*; and the existence of this constant relativity of growth, or constant similarity of form, is of the essence, and may be made the basis of a definition, of the equiangular spiral.[3]

THE GOLDEN TRIANGLE

An application of the gnomon principle which interested the contemporaries of Pythagoras concerns the isosceles triangle *ABC* (Fig. 13.3) which has base angles 72° and apex angle 36°. We met this figure in the "mystical pentagram" (Fig. 2.4), from which we learned that $AB:BC = \phi:1$. Accordingly, we designate the triangles of figure 13.3 the *golden triangles*.

The bisector of $\angle B$ meets *AC* in *D*, so that *D* is the golden cut of *AC*. By this the triangle *ABC* has been divided into two isosceles triangles with equal rights to be called "golden," their apex angles being 36° and 108° and the ratio of their areas $\phi:1$. Bisecting $\angle C$ we obtain *E*, the golden cut of *BD*, and two more golden triangles. This process, producing a series of gnomons, converges to a limiting point *O*, which is the pole of a logarithmic spiral passing successively and in the same order through the three vertices of each of the series of triangles, $\cdots A, B, C, D, \cdots$.

This does not exhaust the intriguing possibilities of the triangle. In addition to the constant recurrence of the golden section, a series obeying the Fibonacci rule, $u_{n+1} = u_n + u_{n-1}$, appears again in figure 13.3.

If we begin with *HG* and call it unit length, then

$$
\begin{aligned}
GF &= 1\phi \\
FE &= 1\phi + 1 \\
ED &= 2\phi + 1 \\
DC &= 3\phi + 2 \\
CB &= 5\phi + 3 \\
BA &= 8\phi + 5
\end{aligned}
$$

Other interesting features are found in figure 13.3. We have bisected the base angles of successive gnomons. If we join the

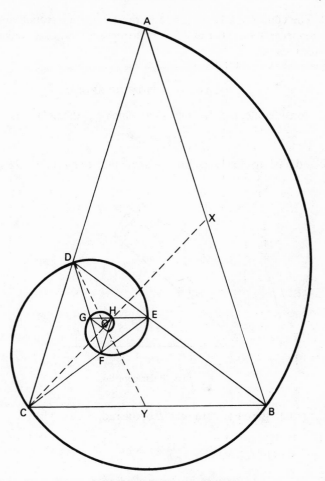

Fig. 13.3. Logarithmic spiral and golden triangles

other base angles to the mid-points of the sides opposite them, e.g. *CX*, *DY*, ···, then

1. The lengths of these medians form a Fibonacci series, and
2. All the medians pass through the pole *O*.

We saw in chapter VII (p. 101) that corresponding points of a series of rectangular gnomons are also the locus of an equiangular spiral.

The connection between the spiral, the golden section and the Fibonacci series is made by referring the spiral to polar co-ordinates.

POLAR EQUATION OF SPIRAL

Consider a curve AB (Fig. 13.4) which has the polar equation

$$r = ae^{b\theta}$$

a and b being constants. Let the angle between a radius OB and a

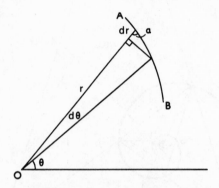

Fig. 13.4. Polar coordinates

tangent to the curve at the end B of the radius be α. Then

$$\frac{dr}{rd\theta} = \cot \alpha$$

From the equation to the curve, $dr/d\theta = abe^{b\theta} = br$, whence $b = \cot \alpha$. Accordingly, α is a constant. Since r increases with θ, we obtain a spiral curve:

$$r = ae^{\theta \cot \alpha}$$

This is the polar equation of an equiangular spiral. The independent variable θ may have any value from $-\infty$ to $+\infty$, so that the curve is unlimited in length.

Using the polar equation, it is a simple matter to make a rough sketch of a portion of the spiral with ruler and compasses only, if

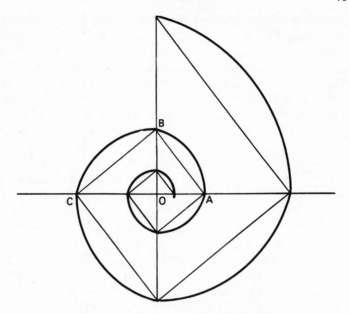

Fig. 13.5. Logarithmic and rectangular spirals

we accept circular arcs as approximations to the actual curve. Alternatively, polar graph paper will give a more accurate result.

Consider three radii separated by right angles (Fig. 13.5),

$$OA = r_1 = ae^{\theta \cot \alpha}$$
$$OB = r_2 = ae^{(\theta + \pi/2) \cot \alpha}$$
$$OC = r_3 = ae^{(\theta + \pi) \cot \alpha}$$

whence $r_2^2 = r_1 r_2$. Thus, OB is mean proportional between OA and OC, whence $\angle ABC$ is a right angle. It follows that the rectangular spiral may serve as the basis of the logarithmic spiral of figure 13.5.

The value of the angle α is at our disposal. An extreme case occurs when $\alpha = 90°$. Then $r_1 = ae^{\theta \cot \alpha}$ degenerates into a circle: $r = a$.

MUSICAL SCALE AND THE SPIRAL

An interesting case involving the equiangular spiral is connected with the musical chromatic scale of 12 semitones. The scale to which a musical instrument is usually tuned is the diatonic scale. In this the frequencies of the note of the common chord $(c - e - g - c')$ are in the simple ratio $4:5:6:8$, the ratios of the corresponding wavelengths being $30:24:20:15$—facts closely related to those that impressed themselves on Pythagoras. The "evenly tempered" scale is slightly different, being a mathematically accurate exponential scale. The musical interval separating the semitones is constant throughout this chromatic scale.

The wavelengths from middle c to c', an octave above it, are shown in the following table, with c' taken as unity; each wavelength is $k = 1.0595$ times that of the semitone above it.

WAVELENGTH $\times$ k

c′	1.0000	0
b	1.0595	15°
a♯	1.1225	30°
a	1.1892	45°
g♯	1.2099	60°
g	1.3348	75°
f♯	1.4141	90°
f	1.4983	105°
e	1.5870	120°
d♯	1.6818	135°
d	1.7819	150°
c♯	1.8878	165°
c	2.0000	180°

If we plot as a smooth curve this tabulation to make a polar graph in which the radii, separated by 15°, are proportional to the wavelengths, we obtain an equiangular (logarithmic) spiral. The ratio of two radii is

$$\frac{r_1}{r_2} = \frac{ae^{\theta_1 \cot \alpha}}{ae^{\theta_2 \cot \alpha}} = e^{(\theta_1 - \theta_2) \cot \alpha}$$

For the octave, $r_1/r_2 = 2$ and $\theta_1 - \theta_2 = 12 \times 15° = 180°$. Hence $2 = e^{\pi \cot \alpha}$ or $\cot \alpha = \log_e 2/\pi$, from which we find that the constant angle α for the equiangular spiral of music is $77°\ 50'$.

By choosing the appropriate value of α, we can ensure that any two radii separated by an angle of $90°$ shall be in the ratio of $\phi : 1$, so that successive segments of the rectangular spiral AB, BC, CD, $\cdots$ obey the Fibonacci rule $u_{n+1} = u_n + u_{n-1}$.

Since a property of the Fibonacci series is that

$$u_{n+1}^2 = u_n \cdot u_{n+2} + (-1)^n \quad \text{(see p. 150)}$$

and we have also

$$r_{n+1}^2 = r_n \cdot r_{n+2}$$

then, when n is large, if $r_n = u_n$, the lengths of successive radii of the equiangular spiral separated by right angles will differ inappreciably from the members of the Fibonacci series.

The condition that $r_2/r_1 = \phi$, when the angle separating the two radii is $\pi/2$, is that

$$\phi = e^{(\pi/2)\cot \alpha}, \qquad \text{or} \qquad \cot \alpha = \frac{2}{\pi} \log_e \phi$$

From this $\alpha = 73°$ approximately. This is the constant angle of the equiangular spiral which incorporates both the golden section and the golden numbers.

◻ ◻ ◻

Before closing this chapter, we must not omit mentioning two more intriguing properties of this remarkable curve. These are easily verified experimentally.

Construct a cardboard template with the outline of an equiangular spiral, piercing a hole at its pole. Then carry out the following simple experiments.

1. Fix the template to a piece of paper which is itself not free to move. Secure one end of a piece of cotton to the template periphery near the pole. Insert a pencil in a small loop at the other end of the cotton. Then, keeping the cotton taut, wrap it around the outline of the spiral. It will be found that the path followed by the pencil is a similar equiangular spiral!

2. Fix a straight edge on a sheet of paper and roll the card-

board spiral along it, marking the successive positions of the
pole. These points lie on a straight line! (Fig. 13.6).

In this chapter we have seen that beauty in nature and beauty
in mathematics are sometimes closely associated. But there is a

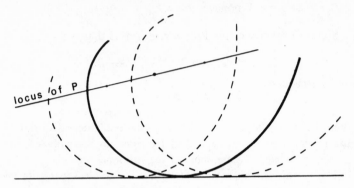

locus of P

Fig. 13.6. Experiment with spiral

difference. Nature's beauty dies. The day dawns when the nautilus
is no more. The rainbow passes, the flower fades, the mountain
crumbles, the star grows cold. But beauty in mathematics—the
divine proportion, the golden rectangle, *spira mirabilis*—endures
for evermore.

APPENDIX

The following excerpts from the writings of CARL G. JUNG *provide an abridged account of his view of the structure of the human psyche.*

I may summarize by observing that we must distinguish three mental levels: (i) consciousness; (ii) the personal unconscious; (iii) the collective unconscious. The personal consists of all those contents which have become unconscious either because, their intensity being lost, they were forgotten, or because consciousness has withdrawn from them, i.e., so-called repression. Finally, this layer contains those elements—partly sense perceptions—which on account of too little intensity have never reached consciousness, and yet in some way have gained access to the psyche. The collective unconscious, being the inheritance of the possibilities of ideas, is not individual but generally human, generally animal even, and represents the real foundation of the individual soul. . . . From the collective unconscious as a timeless universal mind we should expect reactions to the most universal and constant conditions, whether psychological, physiological or physical. From the conscious, on the other hand, we should expect reactions and adaptation phenomena relating to the present; for the conscious is that part of the mind that is preferably limited to events of the moment.[1]

□ □ □

Most of the early impressions in life are soon forgotten and go to form the infantile layer of what I call the personal unconscious. There

177

are definite reasons why I divide the unconscious into two parts. The personal unconscious contains everything forgotten, or repressed, or otherwise subliminal that has been acquired by the individual consciously or unconsciously. Such materials have an unmistakable personal stamp. But other contents are to be found, often enormously strange to the individual and bearing scarcely a trace of personal quality. You may discover such materials frequently in insanity, where they contribute not a little to the confusion and disorientation of the patient. In dreams of normal people such strange contents also occasionally appear.... It is not difficult to define what world these belong to: it is the world of the primitive mind which is deeply unconscious in cultured moderns so long as they are normal, but which rises to the surface when something fatal happens to the conscious. This I call the collective unconscious. "Collective" because it is not an individual acquisition, but rather the functioning of the inherited brain structure, which, in general, is the same in all human beings, and, in certain respects, is the same in all mammals. The inherited brain is the inheritance of the ancient psychic life. It consists of the structural deposits of psychic activities repeated innumerable times in the lives of our ancestors. Our individual conscious is a superstructure upon the collective unconscious, and usually its influence on the conscious is subtle and almost imperceptible. Only at times does it appear in our dreams; and whenever it does, it produces strange and marvellous dreams, remarkable for their beauty, or their demoniacal horror, or for their enigmatic wisdom....[2]

REFERENCES

INTRODUCTION

1. J. Bronowski, *Science and Human Values* (Pelican, 1964) pp. 29–30.
2. *Ibid.*, p. 84–85.
3. C. G. Jung, *Man and his Symbols*, ed. John Freeman (Aldus, 1964).
4. Morris Kline, *Mathematics: a Cultural Approach* (Addison-Wesley, 1962) p. 671.
5. R. Tagore, *The Religion of Man* (Unwin Books) p. 65.

CHAPTER I

1. H. E. Huntley, *The Faith of a Physicist* (Bles, 1960) p. 12.
2. J. Bronowski, *ibid.*, p. 27.
3. C. G. Jung, *The Integration of Personality* (Routledge and Kegan Paul) p. 13.
4. H. J. Eysenck, *Sense and Nonsense in Psychology* (Pelican) p. 319.
5. W. H. Thorpe, *Science, Man and Morals* (Scientific Book Club) p. 88.
6. *Ibid.*, p. 191.
7. H. J. Eysenck, *ibid.*, p. 47.
8. L. P. Jacks, *The Education of the Whole Man* (University of London Press, Portway Reprints).
9. J. Bronowski, *ibid.*, p. 29.

CHAPTER II

1. Martin Gardner, in *Scientific American*, August, 1959, p. 128.

CHAPTER III

1. G. H. Hardy, *A Mathematician's Apology* (Cambridge University Press) p. 63.

2. E. H. Neville, *Proceedings of the London Mathematical Society*, 2nd series XIV, p. 308.

CHAPTER IV

1. Murray Berg, *Fibonacci Quarterly*, Vol. 4, No. 2 (April, 1966), p. 157.

CHAPTER V

1. E. F. Carritt, *The Theory of Beauty*, 4th ed. (Methuen, 1937) p. 6.

CHAPTER VI

1. J. Hadamard, *The Psychology of Invention in the Mathematical Field* (Dover Publications, 1954) p. 127.
2. Quoted by A. Koestler, *The Act of Creation* (Hutchinson) p. 147.
3. G. H. Hardy, *ibid.*, p. 24.
4. G. Polya, *Induction and Analogy in Mathematics* (Princetown) p. 155.

CHAPTER VII

1. H. E. Huntley, *Fibonacci Quarterly*, Vol. 2, No. 2 (April, 1964) p. 104.
2. *Ibid.*, Vol. 2, No. 3 (October, 1964), p. 184.

CHAPTER VIII

1. H. E. Huntley, *The Faith of a Physicist*, p. 62.
2. Geo. Ledin, Jr., *Fibonacci Quarterly*, Vol. 2, No. 4 (Dec., 1964) p. 305.

CHAPTER IX

1. G. H. Hardy, *ibid.*, p. 27.
2. Quoted by Lipton, Matthews, and Rice, *Chess Problems: Introduction to an Art* (Faber and Faber, 1963).

CHAPTER XI

1. H. Poincaré, *Science and Method*, trans. F. Maitland (Dover, 1952) p. 59.

CHAPTER XII

1. Morris Kline, *ibid.*
2. H. E. Licks, *Recreations in Mathematics.*

CHAPTER XIII

1. Crosbie Morrison, *Along the Track* (Whitcombe and Tombs, Melbourne).
2. R. S. Woodworth, *Experimental Psychology* (Methuen) p. 390.
3. D'Arcy Thompson, *On Growth and Form,* abridged edition (Cambridge University Press, 1961) p. 179.

APPENDIX

1. *Symposium on Mind and Earth* (Darmstadt, 1927).
2. *Lecture on Analytical Psychology and Education* (London, 1924).

INDEX

A CATALOGUE OF SELECTED DOVER BOOKS
IN ALL FIELDS OF INTEREST

A CATALOGUE OF SELECTED DOVER BOOKS
IN ALL FIELDS OF INTEREST

AMERICA'S OLD MASTERS, James T. Flexner. Four men emerged unexpectedly from provincial 18th century America to leadership in European art: Benjamin West, J. S. Copley, C. R. Peale, Gilbert Stuart. Brilliant coverage of lives and contributions. Revised, 1967 edition. 69 plates. 365pp. of text.

21806-6 Paperbound $3.00

FIRST FLOWERS OF OUR WILDERNESS: AMERICAN PAINTING, THE COLONIAL PERIOD, James T. Flexner. Painters, and regional painting traditions from earliest Colonial times up to the emergence of Copley, West and Peale Sr., Foster, Gustavus Hesselius, Feke, John Smibert and many anonymous painters in the primitive manner. Engaging presentation, with 162 illustrations. xxii + 368pp.

22180-6 Paperbound $3.50

THE LIGHT OF DISTANT SKIES: AMERICAN PAINTING, 1760-1835, James T. Flexner. The great generation of early American painters goes to Europe to learn and to teach: West, Copley, Gilbert Stuart and others. Allston, Trumbull, Morse; also contemporary American painters—primitives, derivatives, academics—who remained in America. 102 illustrations. xiii + 306pp.

22179-2 Paperbound $3.00

A HISTORY OF THE RISE AND PROGRESS OF THE ARTS OF DESIGN IN THE UNITED STATES, William Dunlap. Much the richest mine of information on early American painters, sculptors, architects, engravers, miniaturists, etc. The only source of information for scores of artists, the major primary source for many others. Unabridged reprint of rare original 1834 edition, with new introduction by James T. Flexner, and 394 new illustrations. Edited by Rita Weiss. 6⅝ x 9⅝.

21695-0, 21696-9, 21697-7 Three volumes, Paperbound $13.50

EPOCHS OF CHINESE AND JAPANESE ART, Ernest F. Fenollosa. From primitive Chinese art to the 20th century, thorough history, explanation of every important art period and form, including Japanese woodcuts; main stress on China and Japan, but Tibet, Korea also included. Still unexcelled for its detailed, rich coverage of cultural background, aesthetic elements, diffusion studies, particularly of the historical period. 2nd, 1913 edition. 242 illustrations. lii + 439pp. of text.

20364-6, 20365-4 Two volumes, Paperbound $6.00

THE GENTLE ART OF MAKING ENEMIES, James A. M. Whistler. Greatest wit of his day deflates Oscar Wilde, Ruskin, Swinburne; strikes back at inane critics, exhibitions, art journalism; aesthetics of impressionist revolution in most striking form. Highly readable classic by great painter. Reproduction of edition designed by Whistler. Introduction by Alfred Werner. xxxvi + 334pp.

21875-9 Paperbound $2.50

INCIDENTS OF TRAVEL IN YUCATAN, John L. Stephens. Classic (1843) exploration of jungles of Yucatan, looking for evidences of Maya civilization. Stephens found many ruins; comments on travel adventures, Mexican and Indian culture. 127 striking illustrations by F. Catherwood. Total of 669 pp.
20926-1, 20927-X Two volumes, Paperbound $5.00

INCIDENTS OF TRAVEL IN CENTRAL AMERICA, CHIAPAS, AND YUCATAN, John L. Stephens. An exciting travel journal and an important classic of archeology. Narrative relates his almost single-handed discovery of the Mayan culture, and exploration of the ruined cities of Copan, Palenque, Utatlan and others; the monuments they dug from the earth, the temples buried in the jungle, the customs of poverty-stricken Indians living a stone's throw from the ruined palaces. 115 drawings by F. Catherwood. Portrait of Stephens. xii + 812pp.
22404-X, 22405-8 Two volumes, Paperbound $6.00

A NEW VOYAGE ROUND THE WORLD, William Dampier. Late 17-century naturalist joined the pirates of the Spanish Main to gather information; remarkably vivid account of buccaneers, pirates; detailed, accurate account of botany, zoology, ethnography of lands visited. Probably the most important early English voyage, enormous implications for British exploration, trade, colonial policy. Also most interesting reading. Argonaut edition, introduction by Sir Albert Gray. New introduction by Percy Adams. 6 plates, 7 illustrations. xlvii + 376pp. 6½ x 9¼.
21900-3 Paperbound $3.00

INTERNATIONAL AIRLINE PHRASE BOOK IN SIX LANGUAGES, Joseph W. Bátor. Important phrases and sentences in English paralleled with French, German, Portuguese, Italian, Spanish equivalents, covering all possible airport-travel situations; created for airline personnel as well as tourist by Language Chief, Pan American Airlines. xiv + 204pp.
22017-6 Paperbound $2.00

STAGE COACH AND TAVERN DAYS, Alice Morse Earle. Detailed, lively account of the early days of taverns; their uses and importance in the social, political and military life; furnishings and decorations; locations; food and drink; tavern signs, etc. Second half covers every aspect of early travel; the roads, coaches, drivers, etc. Nostalgic, charming, packed with fascinating material. 157 illustrations, mostly photographs. xiv + 449pp.
22518-6 Paperbound $4.00

NORSE DISCOVERIES AND EXPLORATIONS IN NORTH AMERICA, Hjalmar R. Holand. The perplexing Kensington Stone, found in Minnesota at the end of the 19th century. Is it a record of a Scandinavian expedition to North America in the 14th century? Or is it one of the most successful hoaxes in history. A scientific detective investigation. Formerly *Westward from Vinland*. 31 photographs, 17 figures. x + 354pp.
22014-1 Paperbound $2.75

A BOOK OF OLD MAPS, compiled and edited by Emerson D. Fite and Archibald Freeman. 74 old maps offer an unusual survey of the discovery, settlement and growth of America down to the close of the Revolutionary war: maps showing Norse settlements in Greenland, the explorations of Columbus, Verrazano, Cabot, Champlain, Joliet, Drake, Hudson, etc., campaigns of Revolutionary war battles, and much more. Each map is accompanied by a brief historical essay. xvi + 299pp. 11 x 13¾.
22084-2 Paperbound $6.00

A History of Costume, Carl Köhler. Definitive history, based on surviving pieces of clothing primarily, and paintings, statues, etc. secondarily. Highly readable text, supplemented by 594 illustrations of costumes of the ancient Mediterranean peoples, Greece and Rome, the Teutonic prehistoric period; costumes of the Middle Ages, Renaissance, Baroque, 18th and 19th centuries. Clear, measured patterns are provided for many clothing articles. Approach is practical throughout. Enlarged by Emma von Sichart. 464pp.   21030-8 Paperbound $3.50

Oriental Rugs, Antique and Modern, Walter A. Hawley. A complete and authoritative treatise on the Oriental rug—where they are made, by whom and how, designs and symbols, characteristics in detail of the six major groups, how to distinguish them and how to buy them. Detailed technical data is provided on periods, weaves, warps, wefts, textures, sides, ends and knots, although no technical background is required for an understanding. 11 color plates, 80 halftones, 4 maps. vi + 320pp. 6⅛ x 9⅛.   22366-3 Paperbound $5.00

Ten Books on Architecture, Vitruvius. By any standards the most important book on architecture ever written. Early Roman discussion of aesthetics of building, construction methods, orders, sites, and every other aspect of architecture has inspired, instructed architecture for about 2,000 years. Stands behind Palladio, Michelangelo, Bramante, Wren, countless others. Definitive Morris H. Morgan translation. 68 illustrations. xii + 331pp.   20645-9 Paperbound $3.50

The Four Books of Architecture, Andrea Palladio. Translated into every major Western European language in the two centuries following its publication in 1570, this has been one of the most influential books in the history of architecture. Complete reprint of the 1738 Isaac Ware edition. New introduction by Adolf Placzek, Columbia Univ. 216 plates. xxii + 110pp. of text. 9½ x 12¾.
                                                    21308-0 Clothbound $10.00

Sticks and Stones: A Study of American Architecture and Civilization, Lewis Mumford.One of the great classics of American cultural history. American architecture from the medieval-inspired earliest forms to the early 20th century; evolution of structure and style, and reciprocal influences on environment. 21 photographic illustrations. 238pp.   20202-X Paperbound $2.00

The American Builder's Companion, Asher Benjamin. The most widely used early 19th century architectural style and source book, for colonial up into Greek Revival periods. Extensive development of geometry of carpentering, construction of sashes, frames, doors, stairs; plans and elevations of domestic and other buildings. Hundreds of thousands of houses were built according to this book, now invaluable to historians, architects, restorers, etc. 1827 edition. 59 plates. 114pp. 7⅞ x 10¾.
                                                    22236-5 Paperbound $3.50

Dutch Houses in the Hudson Valley Before 1776, Helen Wilkinson Reynolds. The standard survey of the Dutch colonial house and outbuildings, with constructional features, decoration, and local history associated with individual homesteads. Introduction by Franklin D. Roosevelt. Map. 150 illustrations. 469pp. 6⅝ x 9¼.   21469-9 Paperbound $4.00

ADVENTURES OF AN AFRICAN SLAVER, Theodore Canot. Edited by Brantz Mayer. A detailed portrayal of slavery and the slave trade, 1820-1840. Canot, an established trader along the African coast, describes the slave economy of the African kingdoms, the treatment of captured negroes, the extensive journeys in the interior to gather slaves, slave revolts and their suppression, harems, bribes, and much more. Full and unabridged republication of 1854 edition. Introduction by Malcom Cowley. 16 illustrations. xvii + 448pp. 22456-2 Paperbound $3.50

MY BONDAGE AND MY FREEDOM, Frederick Douglass. Born and brought up in slavery, Douglass witnessed its horrors and experienced its cruelties, but went on to become one of the most outspoken forces in the American anti-slavery movement. Considered the best of his autobiographies, this book graphically describes the in-human treatment of slaves, its effects on slave owners and slave families, and how Douglass's determination led him to a new life. Unaltered reprint of 1st (1855) edition. xxxii + 464pp. 22457-0 Paperbound $2.50

THE INDIANS' BOOK, recorded and edited by Natalie Curtis. Lore, music, narratives, dozens of drawings by Indians themselves from an authoritative and important survey of native culture among Plains, Southwestern, Lake and Pueblo Indians. Standard work in popular ethnomusicology. 149 songs in full notation. 23 draw-ings, 23 photos. xxxi + 584pp. 6⅝ x 9⅜. 21939-9 Paperbound $4.50

DICTIONARY OF AMERICAN PORTRAITS, edited by Hayward and Blanche Cirker. 4024 portraits of 4000 most important Americans, colonial days to 1905 (with a few important categories, like Presidents, to present). Pioneers, explorers, colonial figures, U. S. officials, politicians, writers, military and naval men, scientists, inven-tors, manufacturers, jurists, actors, historians, educators, notorious figures, Indian chiefs, etc. All authentic contemporary likenesses. The only work of its kind in existence; supplements all biographical sources for libraries. Indispensable to any-one working with American history. 8,000-item classified index, finding lists, other aids. xiv + 756pp. 9¼ x 12¾. 21823-6 Clothbound $30.00

TRITTON'S GUIDE TO BETTER WINE AND BEER MAKING FOR BEGINNERS, S. M. Tritton. All you need to know to make family-sized quantities of over 100 types of grape, fruit, herb and vegetable wines; as well as beers, mead, cider, etc. Com-plete recipes, advice as to equipment, procedures such as fermenting, bottling, and storing wines. Recipes given in British, U. S., and metric measures. Accompanying booklet lists sources in U. S. A. where ingredients may be bought, and additional information. 11 illustrations. 157pp. 5⅝ x 8⅛. (USO) 22090-7 Clothbound $3.50

GARDENING WITH HERBS FOR FLAVOR AND FRAGRANCE, Helen M. Fox. How to grow herbs in your own garden, how to use them in your cooking (over 55 recipes included), legends and myths associated with each species, uses in medicine, per-fumes, etc.—these are elements of one of the few books written especially for Amer-ican herb fanciers. Guides you step-by-step from soil preparation to harvesting and storage for each type of herb. 12 drawings by Louise Mansfield. xiv + 334pp. 22540-2 Paperbound $2.50

VISUAL ILLUSIONS: THEIR CAUSES, CHARACTERISTICS, AND APPLICATIONS, Matthew Luckiesh. Thorough description and discussion of optical illusion, geometric and perspective, particularly; size and shape distortions, illusions of color, of motion; natural illusions; use of illusion in art and magic, industry, etc. Most useful today with op art, also for classical art. Scores of effects illustrated. Introduction by William H. Ittleson. 100 illustrations. xxi + 252pp.
21530-X Paperbound $2.00

A HANDBOOK OF ANATOMY FOR ART STUDENTS, Arthur Thomson. Thorough, virtually exhaustive coverage of skeletal structure, musculature, etc. Full text, supplemented by anatomical diagrams and drawings and by photographs of undraped figures. Unique in its comparison of male and female forms, pointing out differences of contour, texture, form. 211 figures, 40 drawings, 86 photographs. xx + 459pp. 5⅜ x 8⅜.
21163-0 Paperbound $3.50

150 MASTERPIECES OF DRAWING, Selected by Anthony Toney. Full page reproductions of drawings from the early 16th to the end of the 18th century, all beautifully reproduced: Rembrandt, Michelangelo, Dürer, Fragonard, Urs, Graf, Wouwerman, many others. First-rate browsing book, model book for artists. xviii + 150pp. 8⅜ x 11¼.
21032-4 Paperbound $2.50

THE LATER WORK OF AUBREY BEARDSLEY, Aubrey Beardsley. Exotic, erotic, ironic masterpieces in full maturity: Comedy Ballet, Venus and Tannhauser, Pierrot, Lysistrata, Rape of the Lock, Savoy material, Ali Baba, Volpone, etc. This material revolutionized the art world, and is still powerful, fresh, brilliant. With *The Early Work*, all Beardsley's finest work. 174 plates, 2 in color. xiv + 176pp. 8⅛ x 11.
21817-1 Paperbound $3.00

DRAWINGS OF REMBRANDT, Rembrandt van Rijn. Complete reproduction of fabulously rare edition by Lippmann and Hofstede de Groot, completely reedited, updated, improved by Prof. Seymour Slive, Fogg Museum. Portraits, Biblical sketches, landscapes, Oriental types, nudes, episodes from classical mythology—All Rembrandt's fertile genius. Also selection of drawings by his pupils and followers. "Stunning volumes," *Saturday Review*. 550 illustrations. lxxviii + 552pp. 9⅛ x 12¼.
21485-0, 21486-9 Two volumes, Paperbound $10.00

THE DISASTERS OF WAR, Francisco Goya. One of the masterpieces of Western civilization—83 etchings that record Goya's shattering, bitter reaction to the Napoleonic war that swept through Spain after the insurrection of 1808 and to war in general. Reprint of the first edition, with three additional plates from Boston's Museum of Fine Arts. All plates facsimile size. Introduction by Philip Hofer, Fogg Museum. v + 97pp. 9⅜ x 8¼.
21872-4 Paperbound $2.00

GRAPHIC WORKS OF ODILON REDON. Largest collection of Redon's graphic works ever assembled: 172 lithographs, 28 etchings and engravings, 9 drawings. These include some of his most famous works. All the plates from *Odilon Redon: oeuvre graphique complet,* plus additional plates. New introduction and caption translations by Alfred Werner. 209 illustrations. xxvii + 209pp. 9⅛ x 12¼.
21966-8 Paperbound $4.00

DESIGN BY ACCIDENT; A BOOK OF "ACCIDENTAL EFFECTS" FOR ARTISTS AND DESIGNERS, James F. O'Brien. Create your own unique, striking, imaginative effects by "controlled accident" interaction of materials: paints and lacquers, oil and water based paints, splatter, crackling materials, shatter, similar items. Everything you do will be different; first book on this limitless art, so useful to both fine artist and commercial artist. Full instructions. 192 plates showing "accidents," 8 in color. viii + 215pp. 8⅜ x 11¼. 21942-9 Paperbound $3.50

THE BOOK OF SIGNS, Rudolf Koch. Famed German type designer draws 493 beautiful symbols: religious, mystical, alchemical, imperial, property marks, runes, etc. Remarkable fusion of traditional and modern. Good for suggestions of timelessness, smartness, modernity. Text. vi + 104pp. 6⅛ x 9¼.
20162-7 Paperbound $1.25

HISTORY OF INDIAN AND INDONESIAN ART, Ananda K. Coomaraswamy. An unabridged republication of one of the finest books by a great scholar in Eastern art. Rich in descriptive material, history, social backgrounds; Sunga reliefs, Rajput paintings, Gupta temples, Burmese frescoes, textiles, jewelry, sculpture, etc. 400 photos. viii + 423pp. 6⅜ x 9¾. 21436-2 Paperbound $4.00

PRIMITIVE ART, Franz Boas. America's foremost anthropologist surveys textiles, ceramics, woodcarving, basketry, metalwork, etc.; patterns, technology, creation of symbols, style origins. All areas of world, but very full on Northwest Coast Indians. More than 350 illustrations of baskets, boxes, totem poles, weapons, etc. 378 pp.
20025-6 Paperbound $3.00

THE GENTLEMAN AND CABINET MAKER'S DIRECTOR, Thomas Chippendale. Full reprint (third edition, 1762) of most influential furniture book of all time, by master cabinetmaker. 200 plates, illustrating chairs, sofas, mirrors, tables, cabinets, plus 24 photographs of surviving pieces. Biographical introduction by N. Bienenstock. vi + 249pp. 9⅞ x 12¾. 21601-2 Paperbound $4.00

AMERICAN ANTIQUE FURNITURE, Edgar G. Miller, Jr. The basic coverage of all American furniture before 1840. Individual chapters cover type of furniture—clocks, tables, sideboards, etc.—chronologically, with inexhaustible wealth of data. More than 2100 photographs, all identified, commented on. Essential to all early American collectors. Introduction by H. E. Keyes. vi + 1106pp. 7⅞ x 10¾.
21599-7, 21600-4 Two volumes, Paperbound $11.00

PENNSYLVANIA DUTCH AMERICAN FOLK ART, Henry J. Kauffman. 279 photos, 28 drawings of tulipware, Fraktur script, painted tinware, toys, flowered furniture, quilts, samplers, hex signs, house interiors, etc. Full descriptive text. Excellent for tourist, rewarding for designer, collector. Map. 146pp. 7⅞ x 10¾.
21205-X Paperbound $2.50

EARLY NEW ENGLAND GRAVESTONE RUBBINGS, Edmund V. Gillon, Jr. 43 photographs, 226 carefully reproduced rubbings show heavily symbolic, sometimes macabre early gravestones, up to early 19th century. Remarkable early American primitive art, occasionally strikingly beautiful; always powerful. Text. xxvi + 207pp. 8⅜ x 11¼. 21380-3 Paperbound $3.50

How to Know the Wild Flowers, Mrs. William Starr Dana. This is the classical book of American wildflowers (of the Eastern and Central United States), used by hundreds of thousands. Covers over 500 species, arranged in extremely easy to use color and season groups. Full descriptions, much plant lore. This Dover edition is the fullest ever compiled, with tables of nomenclature changes. 174 full-page plates by M. Satterlee. xii + 418pp. 20332-8 Paperbound $2.75

Our Plant Friends and Foes, William Atherton DuPuy. History, economic importance, essential botanical information and peculiarities of 25 common forms of plant life are provided in this book in an entertaining and charming style. Covers food plants (potatoes, apples, beans, wheat, almonds, bananas, etc.), flowers (lily, tulip, etc.), trees (pine, oak, elm, etc.), weeds, poisonous mushrooms and vines, gourds, citrus fruits, cotton, the cactus family, and much more. 108 illustrations. xiv + 290pp. 22272-1 Paperbound $2.50

How to Know the Ferns, Frances T. Parsons. Classic survey of Eastern and Central ferns, arranged according to clear, simple identification key. Excellent introduction to greatly neglected nature area. 57 illustrations and 42 plates. xvi + 215pp. 20740-4 Paperbound $2.00

Manual of the Trees of North America, Charles S. Sargent. America's foremost dendrologist provides the definitive coverage of North American trees and tree-like shrubs. 717 species fully described and illustrated: exact distribution, down to township; full botanical description; economic importance; description of subspecies and races; habitat, growth data; similar material. Necessary to every serious student of tree-life. Nomenclature revised to present. Over 100 locating keys. 783 illustrations. lii + 934pp. 20277-1, 20278-X Two volumes, Paperbound $6.00

Our Northern Shrubs, Harriet L. Keeler. Fine non-technical reference work identifying more than 225 important shrubs of Eastern and Central United States and Canada. Full text covering botanical description, habitat, plant lore, is paralleled with 205 full-page photographs of flowering or fruiting plants. Nomenclature revised by Edward G. Voss. One of few works concerned with shrubs. 205 plates, 35 drawings. xxviii + 521pp. 21989-5 Paperbound $3.75

The Mushroom Handbook, Louis C. C. Krieger. Still the best popular handbook: full descriptions of 259 species, cross references to another 200. Extremely thorough text enables you to identify, know all about any mushroom you are likely to meet in eastern and central U. S. A.: habitat, luminescence, poisonous qualities, use, folklore, etc. 32 color plates show over 50 mushrooms, also 126 other illustrations. Finding keys. vii + 560pp. 21861-9 Paperbound $3.95

Handbook of Birds of Eastern North America, Frank M. Chapman. Still much the best single-volume guide to the birds of Eastern and Central United States. Very full coverage of 675 species, with descriptions, life habits, distribution, similar data. All descriptions keyed to two-page color chart. With this single volume the average birdwatcher needs no other books. 1931 revised edition. 195 illustrations. xxxvi + 581pp. 21489-3 Paperbound $4.50

ALPHABETS AND ORNAMENTS, Ernst Lehner. Well-known pictorial source for decorative alphabets, script examples, cartouches, frames, decorative title pages, calligraphic initials, borders, similar material. 14th to 19th century, mostly European. Useful in almost any graphic arts designing, varied styles. 750 illustrations. 256pp. 7 x 10.   21905-4 Paperbound $4.00

PAINTING: A CREATIVE APPROACH, Norman Colquhoun. For the beginner simple guide provides an instructive approach to painting: major stumbling blocks for beginner; overcoming them, technical points; paints and pigments; oil painting; watercolor and other media and color. New section on "plastic" paints. Glossary. Formerly *Paint Your Own Pictures*. 221pp.   22000-1 Paperbound $1.75

THE ENJOYMENT AND USE OF COLOR, Walter Sargent. Explanation of the relations between colors themselves and between colors in nature and art, including hundreds of little-known facts about color values, intensities, effects of high and low illumination, complementary colors. Many practical hints for painters, references to great masters. 7 color plates, 29 illustrations. x + 274pp.
20944-X Paperbound $2.50

THE NOTEBOOKS OF LEONARDO DA VINCI, compiled and edited by Jean Paul Richter. 1566 extracts from original manuscripts reveal the full range of Leonardo's versatile genius: all his writings on painting, sculpture, architecture, anatomy, astronomy, geography, topography, physiology, mining, music, etc., in both Italian and English, with 186 plates of manuscript pages and more than 500 additional drawings. Includes studies for the Last Supper, the lost Sforza monument, and other works. Total of xlvii + 866pp. 7⅞ x 10¾.
22572-0, 22573-9 Two volumes, Paperbound $10.00

MONTGOMERY WARD CATALOGUE OF 1895. Tea gowns, yards of flannel and pillow-case lace, stereoscopes, books of gospel hymns, the New Improved Singer Sewing Machine, side saddles, milk skimmers, straight-edged razors, high-button shoes, spittoons, and on and on . . . listing some 25,000 items, practically all illustrated. Essential to the shoppers of the 1890's, it is our truest record of the spirit of the period. Unaltered reprint of Issue No. 57, Spring and Summer 1895. Introduction by Boris Emmet. Innumerable illustrations. xiii + 624pp. 8½ x 11⅝.
22377-9 Paperbound $6.95

THE CRYSTAL PALACE EXHIBITION ILLUSTRATED CATALOGUE (LONDON, 1851). One of the wonders of the modern world—the Crystal Palace Exhibition in which all the nations of the civilized world exhibited their achievements in the arts and sciences—presented in an equally important illustrated catalogue. More than 1700 items pictured with accompanying text—ceramics, textiles, cast-iron work, carpets, pianos, sleds, razors, wall-papers, billiard tables, beehives, silverware and hundreds of other artifacts—represent the focal point of Victorian culture in the Western World. Probably the largest collection of Victorian decorative art ever assembled— indispensable for antiquarians and designers. Unabridged republication of the Art-Journal Catalogue of the Great Exhibition of 1851, with all terminal essays. New introduction by John Gloag, F.S.A. xxxiv + 426pp. 9 x 12.
22503-8 Paperbound $4.50

THE PRINCIPLES OF PSYCHOLOGY, William James. The famous long course, complete and unabridged. Stream of thought, time perception, memory, experimental methods—these are only some of the concerns of a work that was years ahead of its time and still valid, interesting, useful. 94 figures. Total of xviii + 1391pp.
20381-6, 20382-4 Two volumes, Paperbound $8.00

THE STRANGE STORY OF THE QUANTUM, Banesh Hoffmann. Non-mathematical but thorough explanation of work of Planck, Einstein, Bohr, Pauli, de Broglie, Schrödinger, Heisenberg, Dirac, Feynman, etc. No technical background needed. "Of books attempting such an account, this is the best," Henry Margenau, Yale. 40-page "Postscript 1959." xii + 285pp.
20518-5 Paperbound $2.00

THE RISE OF THE NEW PHYSICS, A. d'Abro. Most thorough explanation in print of central core of mathematical physics, both classical and modern; from Newton to Dirac and Heisenberg. Both history and exposition; philosophy of science, causality, explanations of higher mathematics, analytical mechanics, electromagnetism, thermodynamics, phase rule, special and general relativity, matrices. No higher mathematics needed to follow exposition, though treatment is elementary to intermediate in level. Recommended to serious student who wishes verbal understanding. 97 illustrations. xvii + 982pp.
20003-5, 20004-3 Two volumes, Paperbound $6.00

GREAT IDEAS OF OPERATIONS RESEARCH, Jagjit Singh. Easily followed non-technical explanation of mathematical tools, aims, results: statistics, linear programming, game theory, queueing theory, Monte Carlo simulation, etc. Uses only elementary mathematics. Many case studies, several analyzed in detail. Clarity, breadth make this excellent for specialist in another field who wishes background. 41 figures. x + 228pp.
21886-4 Paperbound $2.50

GREAT IDEAS OF MODERN MATHEMATICS: THEIR NATURE AND USE, Jagjit Singh. Internationally famous expositor, winner of Unesco's Kalinga Award for science popularization explains verbally such topics as differential equations, matrices, groups, sets, transformations, mathematical logic and other important modern mathematics, as well as use in physics, astrophysics, and similar fields. Superb exposition for layman, scientist in other areas. viii + 312pp.
20587-8 Paperbound $2.50

GREAT IDEAS IN INFORMATION THEORY, LANGUAGE AND CYBERNETICS, Jagjit Singh. The analog and digital computers, how they work, how they are like and unlike the human brain, the men who developed them, their future applications, computer terminology. An essential book for today, even for readers with little math. Some mathematical demonstrations included for more advanced readers. 118 figures. Tables. ix + 338pp.
21694-2 Paperbound $2.50

CHANCE, LUCK AND STATISTICS, Horace C. Levinson. Non-mathematical presentation of fundamentals of probability theory and science of statistics and their applications. Games of chance, betting odds, misuse of statistics, normal and skew distributions, birth rates, stock speculation, insurance. Enlarged edition. Formerly "The Science of Chance." xiii + 357pp.
21007-3 Paperbound $2.50

JIM WHITEWOLF: THE LIFE OF A KIOWA APACHE INDIAN, Charles S. Brant, editor. Spans transition between native life and acculturation period, 1880 on. Kiowa culture, personal life pattern, religion and the supernatural, the Ghost Dance, breakdown in the White Man's world, similar material. 1 map. xii + 144pp.
22015-X Paperbound $1.75

THE NATIVE TRIBES OF CENTRAL AUSTRALIA, Baldwin Spencer and F. J. Gillen. Basic book in anthropology, devoted to full coverage of the Arunta and Warramunga tribes; the source for knowledge about kinship systems, material and social culture, religion, etc. Still unsurpassed. 121 photographs, 89 drawings. xviii + 669pp.
21775-2 Paperbound $5.00

MALAY MAGIC, Walter W. Skeat. Classic (1900); still the definitive work on the folklore and popular religion of the Malay peninsula. Describes marriage rites, birth spirits and ceremonies, medicine, dances, games, war and weapons, etc. Extensive quotes from original sources, many magic charms translated into English. 35 illustrations. Preface by Charles Otto Blagden. xxiv + 685pp.
21760-4 Paperbound $4.00

HEAVENS ON EARTH: UTOPIAN COMMUNITIES IN AMERICA, 1680-1880, Mark Holloway. The finest nontechnical account of American utopias, from the early Woman in the Wilderness, Ephrata, Rappites to the enormous mid 19th-century efflorescence; Shakers, New Harmony, Equity Stores, Fourier's Phalanxes, Oneida, Amana, Fruitlands, etc. "Entertaining and very instructive." *Times Literary Supplement.* 15 illustrations. 246pp.
21593-8 Paperbound $2.00

LONDON LABOUR AND THE LONDON POOR, Henry Mayhew. Earliest (c. 1850) sociological study in English, describing myriad subcultures of London poor. Particularly remarkable for the thousands of pages of direct testimony taken from the lips of London prostitutes, thieves, beggars, street sellers, chimney-sweepers, street-musicians, "mudlarks," "pure-finders," rag-gatherers, "running-patterers," dock laborers, cab-men, and hundreds of others, quoted directly in this massive work. An extraordinarily vital picture of London emerges. 110 illustrations. Total of lxxvi + 1951pp. 6⅝ x 10.
21934-8, 21935-6, 21936-4, 21937-2 Four volumes, Paperbound $14.00

HISTORY OF THE LATER ROMAN EMPIRE, J. B. Bury. Eloquent, detailed reconstruction of Western and Byzantine Roman Empire by a major historian, from the death of Theodosius I (395 A.D.) to the death of Justinian (565). Extensive quotations from contemporary sources; full coverage of important Roman and foreign figures of the time. xxxiv + 965pp. 21829-5 Record, book, album. Monaural. $3.50

AN INTELLECTUAL AND CULTURAL HISTORY OF THE WESTERN WORLD, Harry Elmer Barnes. Monumental study, tracing the development of the accomplishments that make up human culture. Every aspect of man's achievement surveyed from its origins in the Paleolithic to the present day (1964); social structures, ideas, economic systems, art, literature, technology, mathematics, the sciences, medicine, religion, jurisprudence, etc. Evaluations of the contributions of scores of great men. 1964 edition, revised and edited by scholars in the many fields represented. Total of xxix + 1381pp. 21275-0, 21276-9, 21277-7 Three volumes, Paperbound $7.75

THE ARCHITECTURE OF COUNTRY HOUSES, Andrew J. Downing. Together with Vaux's *Villas and Cottages* this is the basic book for Hudson River Gothic architecture of the middle Victorian period. Full, sound discussions of general aspects of housing, architecture, style, decoration, furnishing, together with scores of detailed house plans, illustrations of specific buildings, accompanied by full text. Perhaps the most influential single American architectural book. 1850 edition. Introduction by J. Stewart Johnson. 321 figures, 34 architectural designs. xvi + 560pp.

22003-6 Paperbound $4.00

LOST EXAMPLES OF COLONIAL ARCHITECTURE, John Mead Howells. Full-page photographs of buildings that have disappeared or been so altered as to be denatured, including many designed by major early American architects. 245 plates. xvii + 248pp. 7⅞ x 10¾. 21143-6 Paperbound $3.00

DOMESTIC ARCHITECTURE OF THE AMERICAN COLONIES AND OF THE EARLY REPUBLIC, Fiske Kimball. Foremost architect and restorer of Williamsburg and Monticello covers nearly 200 homes between 1620-1825. Architectural details, construction, style features, special fixtures, floor plans, etc. Generally considered finest work in its area. 219 illustrations of houses, doorways, windows, capital mantels. xx + 314pp. 7⅞ x 10¾. 21743-4 Paperbound $3.50

EARLY AMERICAN ROOMS: 1650-1858, edited by Russell Hawes Kettell. Tour of 12 rooms, each representative of a different era in American history and each furnished, decorated, designed and occupied in the style of the era. 72 plans and elevations, 8-page color section, etc., show fabrics, wall papers, arrangements, etc. Full descriptive text. xvii + 200pp. of text. 8⅜ x 11¼.

21633-0 Paperbound $5.00

THE FITZWILLIAM VIRGINAL BOOK, edited by J. Fuller Maitland and W. B. Squire. Full modern printing of famous early 17th-century ms. volume of 300 works by Morley, Byrd, Bull, Gibbons, etc. For piano or other modern keyboard instrument; easy to read format. xxxvi + 938pp. 8⅜ x 11.

21068-5, 21069-3 Two volumes, Paperbound $8.00

HARPSICHORD MUSIC, Johann Sebastian Bach. Bach Gesellschaft edition. A rich selection of Bach's masterpieces for the harpsichord: the six English Suites, six French Suites, the six Partitas (Clavierübung part I), the Goldberg Variations (Clavierübung part IV), the fifteen Two-Part Inventions and the fifteen Three-Part Sinfonias. Clearly reproduced on large sheets with ample margins; eminently playable. vi + 312pp. 8⅛ x 11. 22360-4 Paperbound $5.00

THE MUSIC OF BACH: AN INTRODUCTION, Charles Sanford Terry. A fine, nontechnical introduction to Bach's music, both instrumental and vocal. Covers organ music, chamber music, passion music, other types. Analyzes themes, developments, innovations. x + 114pp. 21075-8 Paperbound $1.25

BEETHOVEN AND HIS NINE SYMPHONIES, Sir George Grove. Noted British musicologist provides best history, analysis, commentary on symphonies. Very thorough, rigorously accurate; necessary to both advanced student and amateur music lover. 436 musical passages. vii + 407 pp. 20334-4 Paperbound $2.25

TWO LITTLE SAVAGES; BEING THE ADVENTURES OF TWO BOYS WHO LIVED AS INDIANS AND WHAT THEY LEARNED, Ernest Thompson Seton. Great classic of nature and boyhood provides a vast range of woodlore in most palatable form, a genuinely entertaining story. Two farm boys build a teepee in woods and live in it for a month, working out Indian solutions to living problems, star lore, birds and animals, plants, etc. 293 illustrations. vii + 286pp.
20985-7 Paperbound $2.50

PETER PIPER'S PRACTICAL PRINCIPLES OF PLAIN & PERFECT PRONUNCIATION. Alliterative jingles and tongue-twisters of surprising charm, that made their first appearance in America about 1830. Republished in full with the spirited woodcut illustrations from this earliest American edition. 32pp. 4½ x 6⅜.
22560-7 Paperbound $1.00

SCIENCE EXPERIMENTS AND AMUSEMENTS FOR CHILDREN, Charles Vivian. 73 easy experiments, requiring only materials found at home or easily available, such as candles, coins, steel wool, etc.; illustrate basic phenomena like vacuum, simple chemical reaction, etc. All safe. Modern, well-planned. Formerly *Science Games for Children*. 102 photos, numerous drawings. 96pp. 6⅛ x 9¼.
21856-2 Paperbound $1.25

AN INTRODUCTION TO CHESS MOVES AND TACTICS SIMPLY EXPLAINED, Leonard Barden. Informal intermediate introduction, quite strong in explaining reasons for moves. Covers basic material, tactics, important openings, traps, positional play in middle game, end game. Attempts to isolate patterns and recurrent configurations. Formerly *Chess*. 58 figures. 102pp. (USO) 21210-6 Paperbound $1.25

LASKER'S MANUAL OF CHESS, Dr. Emanuel Lasker. Lasker was not only one of the five great World Champions, he was also one of the ablest expositors, theorists, and analysts. In many ways, his Manual, permeated with his philosophy of battle, filled with keen insights, is one of the greatest works ever written on chess. Filled with analyzed games by the great players. A single-volume library that will profit almost any chess player, beginner or master. 308 diagrams. xli x 349pp.
20640-8 Paperbound $2.75

THE MASTER BOOK OF MATHEMATICAL RECREATIONS, Fred Schuh. In opinion of many the finest work ever prepared on mathematical puzzles, stunts, recreations; exhaustively thorough explanations of mathematics involved, analysis of effects, citation of puzzles and games. Mathematics involved is elementary. Translated by F. Göbel. 194 figures. xxiv + 430pp.
22134-2 Paperbound $3.00

MATHEMATICS, MAGIC AND MYSTERY, Martin Gardner. Puzzle editor for Scientific American explains mathematics behind various mystifying tricks: card tricks, stage "mind reading," coin and match tricks, counting out games, geometric dissections, etc. Probability sets, theory of numbers clearly explained. Also provides more than 400 tricks, guaranteed to work, that you can do. 135 illustrations. xii + 176pp.
20338-2 Paperbound $1.50

JOHANN SEBASTIAN BACH, Philipp Spitta. One of the great classics of musicology, this definitive analysis of Bach's music (and life) has never been surpassed. Lucid, nontechnical analyses of hundreds of pieces (30 pages devoted to St. Matthew Passion, 26 to B Minor Mass). Also includes major analysis of 18th-century music. 450 musical examples. 40-page musical supplement. Total of xx + 1799pp.
(EUK) 22278-0, 22279-9 Two volumes, Clothbound $15.00

MOZART AND HIS PIANO CONCERTOS, Cuthbert Girdlestone. The only full-length study of an important area of Mozart's creativity. Provides detailed analyses of all 23 concertos, traces inspirational sources. 417 musical examples. Second edition. 509pp. (USO) 21271-8 Paperbound $3.50

THE PERFECT WAGNERITE: A COMMENTARY ON THE NIBLUNG'S RING, George Bernard Shaw. Brilliant and still relevant criticism in remarkable essays on Wagner's Ring cycle, Shaw's ideas on political and social ideology behind the plots, role of Leitmotifs, vocal requisites, etc. Prefaces. xxi + 136pp.
21707-8 Paperbound $1.50

DON GIOVANNI, W. A. Mozart. Complete libretto, modern English translation; biographies of composer and librettist; accounts of early performances and critical reaction. Lavishly illustrated. All the material you need to understand and appreciate this great work. Dover Opera Guide and Libretto Series; translated and introduced by Ellen Bleiler. 92 illustrations. 209pp.
21134-7 Paperbound $1.50

HIGH FIDELITY SYSTEMS: A LAYMAN'S GUIDE, Roy F. Allison. All the basic information you need for setting up your own audio system: high fidelity and stereo record players, tape records, F.M. Connections, adjusting tone arm, cartridge, checking needle alignment, positioning speakers, phasing speakers, adjusting hums, trouble-shooting, maintenance, and similar topics. Enlarged 1965 edition. More than 50 charts, diagrams, photos. iv + 91pp. 21514-8 Paperbound $1.25

REPRODUCTION OF SOUND, Edgar Villchur. Thorough coverage for laymen of high fidelity systems, reproducing systems in general, needles, amplifiers, preamps, loudspeakers, feedback, explaining physical background. "A rare talent for making technicalities vividly comprehensible," R. Darrell, High Fidelity. 69 figures. iv + 92pp. 21515-6 Paperbound $1.25

HEAR ME TALKIN' TO YA: THE STORY OF JAZZ AS TOLD BY THE MEN WHO MADE IT, Nat Shapiro and Nat Hentoff. Louis Armstrong, Fats Waller, Jo Jones, Clarence Williams, Billy Holiday, Duke Ellington, Jelly Roll Morton and dozens of other jazz greats tell how it was in Chicago's South Side, New Orleans, depression Harlem and the modern West Coast as jazz was born and grew. xvi + 429pp.
21726-4 Paperbound $2.50

FABLES OF AESOP, translated by Sir Roger L'Estrange. A reproduction of the very rare 1931 Paris edition; a selection of the most interesting fables, together with 50 imaginative drawings by Alexander Calder. v + 128pp. 6½x9¼.
21780-9 Paperbound $1.50

THE RED FAIRY BOOK, Andrew Lang. Lang's color fairy books have long been children's favorites. This volume includes Rapunzel, Jack and the Bean-stalk and 35 other stories, familiar and unfamiliar. 4 plates, 93 illustrations x + 367pp.
21673-X Paperbound $2.50

THE BLUE FAIRY BOOK, Andrew Lang. Lang's tales come from all countries and all times. Here are 37 tales from Grimm, the Arabian Nights, Greek Mythology, and other fascinating sources. 8 plates, 130 illustrations. xi + 390pp.
21437-0 Paperbound $2.50

HOUSEHOLD STORIES BY THE BROTHERS GRIMM. Classic English-language edition of the well-known tales — Rumpelstiltskin, Snow White, Hansel and Gretel, The Twelve Brothers, Faithful John, Rapunzel, Tom Thumb (52 stories in all). Translated into simple, straightforward English by Lucy Crane. Ornamented with head-pieces, vignettes, elaborate decorative initials and a dozen full-page illustrations by Walter Crane. x + 269pp.
21080-4 Paperbound $2.50

THE MERRY ADVENTURES OF ROBIN HOOD, Howard Pyle. The finest modern versions of the traditional ballads and tales about the great English outlaw. Howard Pyle's complete prose version, with every word, every illustration of the first edition. Do not confuse this facsimile of the original (1883) with modern editions that change text or illustrations. 23 plates plus many page decorations. xxii + 296pp.
22043-5 Paperbound $2.50

THE STORY OF KING ARTHUR AND HIS KNIGHTS, Howard Pyle. The finest children's version of the life of King Arthur; brilliantly retold by Pyle, with 48 of his most imaginative illustrations. xviii + 313pp. 6⅛ x 9¼.
21445-1 Paperbound $2.50

THE WONDERFUL WIZARD OF OZ, L. Frank Baum. America's finest children's book in facsimile of first edition with all Denslow illustrations in full color. The edition a child should have. Introduction by Martin Gardner. 23 color plates, scores of drawings. iv + 267pp. 20691-2 Paperbound $2.50

THE MARVELOUS LAND OF OZ, L. Frank Baum. The second Oz book, every bit as imaginative as the Wizard. The hero is a boy named Tip, but the Scarecrow and the Tin Woodman are back, as is the Oz magic. 16 color plates, 120 drawings by John R. Neill. 287pp. 20692-0 Paperbound $2.50

THE MAGICAL MONARCH OF MO, L. Frank Baum. Remarkable adventures in a land even stranger than Oz. The best of Baum's books not in the Oz series. 15 color plates and dozens of drawings by Frank Verbeck. xviii + 237pp.
21892-9 Paperbound $2.25

THE BAD CHILD'S BOOK OF BEASTS, MORE BEASTS FOR WORSE CHILDREN, A MORAL ALPHABET, Hilaire Belloc. Three complete humor classics in one volume. Be kind to the frog, and do not call him names . . . and 28 other whimsical animals. Familiar favorites and some not so well known. Illustrated by Basil Blackwell. 156pp. (USO) 20749-8 Paperbound $1.50

AMERICAN FOOD AND GAME FISHES, David S. Jordan and Barton W. Evermann. Definitive source of information, detailed and accurate enough to enable the sportsman and nature lover to identify conclusively some 1,000 species and sub-species of North American fish, sought for food or sport. Coverage of range, physiology, habits, life history, food value. Best methods of capture, interest to the angler, advice on bait, fly-fishing, etc. 338 drawings and photographs. 1 + 574pp. 6⅝ x 9⅜.
22383-1 Paperbound $4.50

THE FROG BOOK, Mary C. Dickerson. Complete with extensive finding keys, over 300 photographs, and an introduction to the general biology of frogs and toads, this is the classic non-technical study of Northeastern and Central species. 58 species; 290 photographs and 16 color plates. xvii + 253pp.
21973-9 Paperbound $4.00

THE MOTH BOOK: A GUIDE TO THE MOTHS OF NORTH AMERICA, William J. Holland. Classical study, eagerly sought after and used for the past 60 years. Clear identification manual to more than 2,000 different moths, largest manual in existence. General information about moths, capturing, mounting, classifying, etc., followed by species by species descriptions. 263 illustrations plus 48 color plates show almost every species, full size. 1968 edition, preface, nomenclature changes by A. E. Brower. xxiv + 479pp. of text. 6½ x 9¼.
21948-8 Paperbound $5.00

THE SEA-BEACH AT EBB-TIDE, Augusta Foote Arnold. Interested amateur can identify hundreds of marine plants and animals on coasts of North America; marine algae; seaweeds; squids; hermit crabs; horse shoe crabs; shrimps; corals; sea anemones; etc. Species descriptions cover: structure; food; reproductive cycle; size; shape; color; habitat; etc. Over 600 drawings. 85 plates. xii + 490pp.
21949-6 Paperbound $3.50

COMMON BIRD SONGS, Donald J. Borror. 33⅓ 12-inch record presents songs of 60 important birds of the eastern United States. A thorough, serious record which provides several examples for each bird, showing different types of song, individual variations, etc. Inestimable identification aid for birdwatcher. 32-page booklet gives text about birds and songs, with illustration for each bird.
21829-5 Record, book, album. Monaural. $2.75

FADS AND FALLACIES IN THE NAME OF SCIENCE, Martin Gardner. Fair, witty appraisal of cranks and quacks of science: Atlantis, Lemuria, hollow earth, flat earth, Velikovsky, orgone energy, Dianetics, flying saucers, Bridey Murphy, food fads, medical fads, perpetual motion, etc. Formerly "In the Name of Science." x + 363pp.
20394-8 Paperbound $2.00

HOAXES, Curtis D. MacDougall. Exhaustive, unbelievably rich account of great hoaxes: Locke's moon hoax, Shakespearean forgeries, sea serpents, Loch Ness monster, Cardiff giant, John Wilkes Booth's mummy, Disumbrationist school of art, dozens more; also journalism, psychology of hoaxing. 54 illustrations. xi + 338pp.
20465-0 Paperbound $2.75

POEMS OF ANNE BRADSTREET, edited with an introduction by Robert Hutchinson. A new selection of poems by America's first poet and perhaps the first significant woman poet in the English language. 48 poems display her development in works of considerable variety—love poems, domestic poems, religious meditations, formal elegies, "quaternions," etc. Notes, bibliography. viii + 222pp.
22160-1 Paperbound $2.00

THREE GOTHIC NOVELS: THE CASTLE OF OTRANTO BY HORACE WALPOLE; VATHEK BY WILLIAM BECKFORD; THE VAMPYRE BY JOHN POLIDORI, WITH FRAGMENT OF A NOVEL BY LORD BYRON, edited by E. F. Bleiler. The first Gothic novel, by Walpole; the finest Oriental tale in English, by Beckford; powerful Romantic supernatural story in versions by Polidori and Byron. All extremely important in history of literature; all still exciting, packed with supernatural thrills, ghosts, haunted castles, magic, etc. xl + 291pp.
21232-7 Paperbound $2.00

THE BEST TALES OF HOFFMANN, E. T. A. Hoffmann. 10 of Hoffmann's most important stories, in modern re-editings of standard translations: Nutcracker and the King of Mice, Signor Formica, Automata, The Sandman, Rath Krespel, The Golden Flowerpot, Master Martin the Cooper, The Mines of Falun, The King's Betrothed, A New Year's Eve Adventure. 7 illustrations by Hoffmann. Edited by E. F. Bleiler. xxxix + 419pp.
21793-0 Paperbound $2.50

GHOST AND HORROR STORIES OF AMBROSE BIERCE, Ambrose Bierce. 23 strikingly modern stories of the horrors latent in the human mind: The Eyes of the Panther, The Damned Thing, An Occurrence at Owl Creek Bridge, An Inhabitant of Carcosa, etc., plus the dream-essay, Visions of the Night. Edited by E. F. Bleiler. xxii + 199pp.
20767-6 Paperbound $1.50

BEST GHOST STORIES OF J. S. LeFANU, J. Sheridan LeFanu. Finest stories by Victorian master often considered greatest supernatural writer of all. Carmilla, Green Tea, The Haunted Baronet, The Familiar, and 12 others. Most never before available in the U. S. A. Edited by E. F. Bleiler. 8 illustrations from Victorian publications. xvii + 467pp.
20415-4 Paperbound $2.50

THE TIME STREAM, THE GREATEST ADVENTURE, AND THE PURPLE SAPPHIRE—THREE SCIENCE FICTION NOVELS, John Taine (Eric Temple Bell). Great American mathematician was also foremost science fiction novelist of the 1920's. *The Time Stream,* one of all-time classics, uses concepts of circular time; *The Greatest Adventure,* incredibly ancient biological experiments from Antarctica threaten to escape; *The Purple Sapphire,* superscience, lost races in Central Tibet, survivors of the Great Race. 4 illustrations by Frank R. Paul. v + 532pp.
21180-0 Paperbound $3.00

SEVEN SCIENCE FICTION NOVELS, H. G. Wells. The standard collection of the great novels. Complete, unabridged. *First Men in the Moon, Island of Dr. Moreau, War of the Worlds, Food of the Gods, Invisible Man, Time Machine, In the Days of the Comet.* Not only science fiction fans, but every educated person owes it to himself to read these novels. 1015pp.
20264-X Clothbound $5.00

THE PHILOSOPHY OF THE UPANISHADS, Paul Deussen. Clear, detailed statement of upanishadic system of thought, generally considered among best available. History of these works, full exposition of system emergent from them, parallel concepts in the West. Translated by A. S. Geden. xiv + 429pp.
21616-0 Paperbound $3.00

LANGUAGE, TRUTH AND LOGIC, Alfred J. Ayer. Famous, remarkably clear introduction to the Vienna and Cambridge schools of Logical Positivism; function of philosophy, elimination of metaphysical thought, nature of analysis, similar topics. "Wish I had written it myself," Bertrand Russell. 2nd, 1946 edition. 160pp.
20010-8 Paperbound $1.35

THE GUIDE FOR THE PERPLEXED, Moses Maimonides. Great classic of medieval Judaism, major attempt to reconcile revealed religion (Pentateuch, commentaries) and Aristotelian philosophy. Enormously important in all Western thought. Unabridged Friedländer translation. 50-page introduction. lix + 414pp.
(USO) 20351-4 Paperbound $2.50

OCCULT AND SUPERNATURAL PHENOMENA, D. H. Rawcliffe. Full, serious study of the most persistent delusions of mankind: crystal gazing, mediumistic trance, stigmata, lycanthropy, fire walking, dowsing, telepathy, ghosts, ESP, etc., and their relation to common forms of abnormal psychology. Formerly *Illusions and Delusions of the Supernatural and the Occult.* iii + 551pp. 20503-7 Paperbound $3.50

THE EGYPTIAN BOOK OF THE DEAD: THE PAPYRUS OF ANI, E. A. Wallis Budge. Full hieroglyphic text, interlinear transliteration of sounds, word for word translation, then smooth, connected translation; Theban recension. Basic work in Ancient Egyptian civilization; now even more significant than ever for historical importance, dilation of consciousness, etc. clvi + 377pp. 6½ x 9¼.
21866-X Paperbound $3.95

PSYCHOLOGY OF MUSIC, Carl E. Seashore. Basic, thorough survey of everything known about psychology of music up to 1940's; essential reading for psychologists, musicologists. Physical acoustics; auditory apparatus; relationship of physical sound to perceived sound; role of the mind in sorting, altering, suppressing, creating sound sensations; musical learning, testing for ability, absolute pitch, other topics. Records of Caruso, Menuhin analyzed. 88 figures. xix + 408pp.
21851-1 Paperbound $2.75

THE I CHING (THE BOOK OF CHANGES), translated by James Legge. Complete translated text plus appendices by Confucius, of perhaps the most penetrating divination book ever compiled. Indispensable to all study of early Oriental civilizations. 3 plates. xxiii + 448pp. 21062-6 Paperbound $3.00

THE UPANISHADS, translated by Max Müller. Twelve classical upanishads: Chandogya, Kena, Aitareya, Kaushitaki, Isa, Katha, Mundaka, Taittiriyaka, Brhadaranyaka, Svetasvatara, Prasna, Maitriyana. 160-page introduction, analysis by Prof. Müller. Total of 826pp. 20398-0, 20399-9 Two volumes, Paperbound $5.00

PLANETS, STARS AND GALAXIES: DESCRIPTIVE ASTRONOMY FOR BEGINNERS, A. E. Fanning. Comprehensive introductory survey of astronomy: the sun, solar system, stars, galaxies, universe, cosmology; up-to-date, including quasars, radio stars, etc. Preface by Prof. Donald Menzel. 24pp. of photographs. 189pp. 5¼ x 8¼.
21680-2 Paperbound $1.50

TEACH YOURSELF CALCULUS, P. Abbott. With a good background in algebra and trig, you can teach yourself calculus with this book. Simple, straightforward introduction to functions of all kinds, integration, differentiation, series, etc. "Students who are beginning to study calculus method will derive great help from this book." Faraday House Journal. 308pp. 20683-1 Clothbound $2.00

TEACH YOURSELF TRIGONOMETRY, P. Abbott. Geometrical foundations, indices and logarithms, ratios, angles, circular measure, etc. are presented in this sound, easy-to-use text. Excellent for the beginner or as a brush up, this text carries the student through the solution of triangles. 204pp. 20682-3 Clothbound $2.00

TEACH YOURSELF ANATOMY, David LeVay. Accurate, inclusive, profusely illustrated account of structure, skeleton, abdomen, muscles, nervous system, glands, brain, reproductive organs, evolution. "Quite the best and most readable account,' *Medical Officer.* 12 color plates. 164 figures. 311pp. 4¾ x 7.
21651-9 Clothbound $2.50

TEACH YOURSELF PHYSIOLOGY, David LeVay. Anatomical, biochemical bases; digestive, nervous, endocrine systems; metabolism; respiration; muscle; excretion; temperature control; reproduction. "Good elementary exposition," *The Lancet.* 6 color plates. 44 illustrations. 208pp. 4¼ x 7. 21658-6 Clothbound $2.50

THE FRIENDLY STARS, Martha Evans Martin. Classic has taught naked-eye observation of stars, planets to hundreds of thousands, still not surpassed for charm, lucidity, adequacy. Completely updated by Professor Donald H. Menzel, Harvard Observatory. 25 illustrations. 16 x 30 chart. x + 147pp. 21099-5 Paperbound $1.25

MUSIC OF THE SPHERES: THE MATERIAL UNIVERSE FROM ATOM TO QUASAR, SIMPLY EXPLAINED, Guy Murchie. Extremely broad, brilliantly written popular account begins with the solar system and reaches to dividing line between matter and nonmatter; latest understandings presented with exceptional clarity. Volume One: Planets, stars, galaxies, cosmology, geology, celestial mechanics, latest astronomical discoveries; Volume Two: Matter, atoms, waves, radiation, relativity, chemical action, heat, nuclear energy, quantum theory, music, light, color, probability, antimatter, antigravity, and similar topics. 319 figures. 1967 (second) edition. Total of xx + 644pp. 21809-0, 21810-4 Two volumes, Paperbound $5.00

OLD-TIME SCHOOLS AND SCHOOL BOOKS, Clifton Johnson. Illustrations and rhymes from early primers, abundant quotations from early textbooks, many anecdotes of school life enliven this study of elementary schools from Puritans to middle 19th century. Introduction by Carl Withers. 234 illustrations. xxxiii + 381pp.
21031-6 Paperbound $2.50

EAST O' THE SUN AND WEST O' THE MOON, George W. Dasent. Considered the best of all translations of these Norwegian folk tales, this collection has been enjoyed by generations of children (and folklorists too). Includes True and Untrue, Why the Sea is Salt, East O' the Sun and West O' the Moon, Why the Bear is Stumpy-Tailed, Boots and the Troll, The Cock and the Hen, Rich Peter the Pedlar, and 52 more. The only edition with all 59 tales. 77 illustrations by Erik Werenskiold and Theodor Kittelsen. xv + 418pp. 22521-6 Paperbound $3.50

GOOPS AND HOW TO BE THEM, Gelett Burgess. Classic of tongue-in-cheek humor, masquerading as etiquette book. 87 verses, twice as many cartoons, show mischievous Goops as they demonstrate to children virtues of table manners, neatness, courtesy, etc. Favorite for generations. viii + 88pp. 6½ x 9¼.
22233-0 Paperbound $1.25

ALICE'S ADVENTURES UNDER GROUND, Lewis Carroll. The first version, quite different from the final *Alice in Wonderland*, printed out by Carroll himself with his own illustrations. Complete facsimile of the "million dollar" manuscript Carroll gave to Alice Liddell in 1864. Introduction by Martin Gardner. viii + 96pp. Title and dedication pages in color. 21482-6 Paperbound $1.25

THE BROWNIES, THEIR BOOK, Palmer Cox. Small as mice, cunning as foxes, exuberant and full of mischief, the Brownies go to the zoo, toy shop, seashore, circus, etc., in 24 verse adventures and 266 illustrations. Long a favorite, since their first appearance in St. Nicholas Magazine. xi + 144pp. 6⅝ x 9¼.
21265-3 Paperbound $1.75

SONGS OF CHILDHOOD, Walter De La Mare. Published (under the pseudonym Walter Ramal) when De La Mare was only 29, this charming collection has long been a favorite children's book. A facsimile of the first edition in paper, the 47 poems capture the simplicity of the nursery rhyme and the ballad, including such lyrics as I Met Eve, Tartary, The Silver Penny. vii + 106pp. 21972-0 Paperbound $1.25

THE COMPLETE NONSENSE OF EDWARD LEAR, Edward Lear. The finest 19th-century humorist-cartoonist in full: all nonsense limericks, zany alphabets, Owl and Pussycat, songs, nonsense botany, and more than 500 illustrations by Lear himself. Edited by Holbrook Jackson. xxix + 287pp. (USO) 20167-8 Paperbound $2.00

BILLY WHISKERS: THE AUTOBIOGRAPHY OF A GOAT, Frances Trego Montgomery. A favorite of children since the early 20th century, here are the escapades of that rambunctious, irresistible and mischievous goat—Billy Whiskers. Much in the spirit of *Peck's Bad Boy*, this is a book that children never tire of reading or hearing. All the original familiar illustrations by W. H. Fry are included: 6 color plates, 18 black and white drawings. 159pp. 22345-0 Paperbound $2.00

MOTHER GOOSE MELODIES. Faithful republication of the fabulously rare Munroe and Francis "copyright 1833" Boston edition—the most important Mother Goose collection, usually referred to as the "original." Familiar rhymes plus many rare ones, with wonderful old woodcut illustrations. Edited by E. F. Bleiler. 128pp. 4½ x 6⅜. 22577-1 Paperbound $1.25

MATHEMATICAL PUZZLES FOR BEGINNERS AND ENTHUSIASTS, Geoffrey Mott-Smith. 189 puzzles from easy to difficult—involving arithmetic, logic, algebra, properties of digits, probability, etc.—for enjoyment and mental stimulus. Explanation of mathematical principles behind the puzzles. 135 illustrations. viii + 248pp.
20198-8 Paperbound $1.75

PAPER FOLDING FOR BEGINNERS, William D. Murray and Francis J. Rigney. Easiest book on the market, clearest instructions on making interesting, beautiful origami. Sail boats, cups, roosters, frogs that move legs, bonbon boxes, standing birds, etc. 40 projects; more than 275 diagrams and photographs. 94pp.
20713-7 Paperbound $1.00

TRICKS AND GAMES ON THE POOL TABLE, Fred Herrmann. 79 tricks and games— some solitaires, some for two or more players, some competitive games—to entertain you between formal games. Mystifying shots and throws, unusual caroms, tricks involving such props as cork, coins, a hat, etc. Formerly *Fun on the Pool Table*. 77 figures. 95pp.
21814-7 Paperbound $1.00

HAND SHADOWS TO BE THROWN UPON THE WALL: A SERIES OF NOVEL AND AMUSING FIGURES FORMED BY THE HAND, Henry Bursill. Delightful picturebook from great-grandfather's day shows how to make 18 different hand shadows: a bird that flies, duck that quacks, dog that wags his tail, camel, goose, deer, boy, turtle, etc. Only book of its sort. vi + 33pp. 6½ x 9¼.   21779-5 Paperbound $1.00

WHITTLING AND WOODCARVING, E. J. Tangerman. 18th printing of best book on market. "If you can cut a potato you can carve" toys and puzzles, chains, chessmen, caricatures, masks, frames, woodcut blocks, surface patterns, much more. Information on tools, woods, techniques. Also goes into serious wood sculpture from Middle Ages to present, East and West. 464 photos, figures. x + 293pp.
20965-2 Paperbound $2.00

HISTORY OF PHILOSOPHY, Julián Marias. Possibly the clearest, most easily followed, best planned, most useful one-volume history of philosophy on the market; neither skimpy nor overfull. Full details on system of every major philosopher and dozens of less important thinkers from pre-Socratics up to Existentialism and later. Strong on many European figures usually omitted. Has gone through dozens of editions in Europe. 1966 edition, translated by Stanley Appelbaum and Clarence Strowbridge. xviii + 505pp.   21739-6 Paperbound $3.00

YOGA: A SCIENTIFIC EVALUATION, Kovoor T. Behanan. Scientific but non-technical study of physiological results of yoga exercises; done under auspices of Yale U. Relations to Indian thought, to psychoanalysis, etc. 16 photos. xxiii + 270pp.
20505-3 Paperbound $2.50

Prices subject to change without notice.
Available at your book dealer or write for free catalogue to Dept. GI, Dover Publications, Inc., 180 Varick St., N. Y., N. Y. 10014. Dover publishes more than 150 books each year on science, elementary and advanced mathematics, biology, music, art, literary history, social sciences and other areas.